Encyclopaedia of

World Geography

Encyclopaedia of
World Geography

Vol. - 3

Chandra Vijay Purty

ABD PUBLISHERS
Jaipur ■ New Delhi

ABD PUBLISHERS

Regd. Off. "Bony Residency", Gate No. 2
Opp. Tilak Public School, Vishweswariya Nagar,
Gopalpura Road, Jaipur - 302018, India
Ph: 0141 - 2761280, Telefax: 2761381
Email: abdpublishersjpr@gmail.com

Delhi Office: 102, 1st Floor, Satyam House,
4327/3, Ansari Road, Darya Ganj,
New Delhi - 110002, India
Ph: 011-45652440

visit us: www.oxfordbookcompany.com

Published in India, 2024

ISBN : 9788183763127 (7 Vols. Set)

Typeset by Shivangi Computer, Jaipur

Printed and bound in India by Thomson Press Ltd.,
New Delhi

Preface

Geography is the study of the Earth and its lands, features, inhabitants, and phenomena. A literal translation would be "to describe or write about the Earth". The first person to use the word "geography" was Eratosthenes (276-194 B.C.). Four historical traditions in geographical research are the spatial analysis of natural and human phenomena (geography as a study of distribution), area studies (places and regions), study of man-land relationship, and research in earth sciences. Nonetheless, modern geography is an all-encompassing discipline that foremost seeks to understand the Earth and all of its human and natural complexities—not merely where objects are, but how they have changed and come to be. As "the bridge between the human and physical sciences," geography is divided into two main branches—human geography and physical geography.

The oldest known world maps date back to ancient Babylon from the 9th century BC. The best known Babylonian world map, however, is the *Imago Mundi* of 600 BC. The map as reconstructed by Eckhard Unger shows Babylon on the Euphrates, surrounded by a circular landmass showing Assyria, Urartu and several cities, in turn surrounded by a "bitter river" (Oceanus), with seven islands arranged around it so as to form a seven-pointed star. The accompanying text mentions seven outer regions beyond the encircling ocean. The descriptions of five of them have survived. In contrast to the *Imago Mundi*, an earlier Babylonian world map dating back to the 9th century BC depicted Babylon as being further north from the centre of the world, though it is not certain what that centre was supposed to represent.

The ideas of Anaximander (c. 610 B.C.-c. 545 B.C.), considered by later Greek writers to be the true founder of geography, come to us through fragments quoted by his successors. Anaximander is credited with the invention of the gnomon, the simple yet efficient Greek instrument that allowed the early measurement of latitude. Thales, Anaximander is also credited with the prediction of eclipses. The foundations of geography can be traced to the ancient cultures, such as the ancient, medieval, and early modern Chinese. The Greeks, who were the first to explore geography as both art and science, achieved this through Cartography, Philosophy, and Literature, or through Mathematics. There is some debate about who was the first person to assert that the Earth is spherical in shape, with the credit going either to Parmenides or Pythagoras. Anaxagoras was able to demonstrate that the profile of the Earth was circular by explaining eclipses. However, he still believed that the Earth was a flat disk, as did many of his contemporaries. One of the first estimates of the radius of the Earth was made by Eratosthenes.

The first rigorous system of latitude and longitude lines is credited to Hipparchus. He employed a sexagesimal system that was derived from Babylonian mathematics. The parallels and meridians were sub-divided into 360°, with each degree further subdivided 602 (minutes). To measure the longitude at different location on Earth, he suggested using eclipses to determine the relative difference in time. The extensive mapping by the Romans as they explored new lands would later provide a high level of information for Ptolemy to construct detailed atlases. He extended the work of Hipparchus, using a grid system on his maps and adopting a length of 56.5 miles for a degree.

From the 3rd century onwards, Chinese methods of geographical study and writing of geographical literature became much more complex than what was found in Europe at the time (until the 13th century). Chinese geographers such as Liu An, Pei Xiu, Jia Dan, Shen Kuo, Fan Chengda, Zhou Daguan, and Xu Xiake wrote important treatises, yet by the 17th century, advanced ideas and methods of Western-style geography were adopted in China.

This book contains the fundamental and basic information of the subject.

Chandra Vijay Purty

Contents

1

Switzerland: The Heat of the Alps

Switzerland (German: die Schweiz, French: la Suisse, Italian: Svizzera, Romansh: la Svizra), officially the Swiss Confederation (Confœderatio Helvetica in Latin, hence its ISO country codes CH and CHE), is a federal republic consisting of 26 cantons, with Bern as the seat of the federal authorities. The country is situated in Western Europe where it is bordered by Germany to the north, France to the west, Italy to the south, and Austria and Liechtenstein to the east.

Switzerland is a landlocked country whose territory is geographically divided between the Alps, the Central Plateau and the Jura that yields a total area of 41,285 km2 (15,940 sq mi). The Swiss population of approximately 7.8 million people concentrates mostly on the Plateau, where the largest cities are to be found. Among them are the two global cities and economic centres of Zurich and Geneva. Switzerland is one of the richest countries in the world by per capita gross domestic product, with a nominal per capita GDP of $67,384. Zurich and Geneva have respectively been ranked as the cities with the second and third highest quality of life in the world.

The Swiss Confederation has a long history of neutrality—it has not been in a state of war internationally since 1815—and did not join the United Nations until 2002. Switzerland is home to many international organisations, including the World Economic Forum, the International Olympic Committee, the Red Cross, the World Trade Organization, FIFA, and the second largest UN office. On the European level it was a founder of the European Free Trade Association and is part of the Schengen Agreement-although it is notably not a member of the European Union, nor the European Economic Area.

Switzerland comprises three main linguistic and cultural regions: German, French, and Italian, to which the Romansh-speaking valleys are added. The Swiss therefore do not form a nation in the sense of a common ethnic or linguistic identity. The strong sense of belonging to the country is founded on the common historical background, shared values (federalism, direct democracy, neutrality) and Alpine symbolism. The establishment of the Swiss Confederation is traditionally dated to 1 August 1291; Swiss National Day is celebrated on the anniversary.

Etymology

The English name Switzerland is a compound containing Switzer, an obsolete term for the Swiss, in use during the 16th to 19th centuries. The English adjective Swiss is a loan from French Suisse, also in use since the 16th century. The name Switzer is from the Alemannic Schwiizer, in origin an inhabitant of Schwyz and its associated territory, one of the Waldstatten cantons which formed the nucleus of the Old Swiss Confederacy. The toponym itself is first attested in 972, as Old High German Suittes, ultimately perhaps related to suedan "to burn", referring to the area of forest that was burned and cleared to build. The name was extended to the area dominated by the canton, and after the Swabian War of 1499 gradually came to be used for the entire Confederation.

The Swiss German name of the country, Schwiiz, is homophonous to that of the canton and the settlement, but distinguished by the use of the definite article (d'Schwiiz for the Confederation, but simply Schwyz for the canton and the town).

The Neo-Latin name Confoederatio Helvetica was introduced gradually after the formation of the federal state in 1848, harking back to the Napoleonic Helvetic Republic, appearing on coins from 1879, inscribed on the Federal Palace in 1902 and after 1948 used in the official seal.. It is derived from the name of the Helvetii, a Gaulish tribe living on the Swiss plateau before the Roman era. Helvetia appears as a national personification of the Swiss confederacy in the 17th century, with a 1672 play by Johann Caspar Weissenbach.

History

Switzerland has existed as a state in its present form since the adoption of the Swiss Federal Constitution in 1848. The precursors of modern Switzerland established a protective alliance at the end of the 13th century, forming a loose confederation of states which persisted for centuries.

Early History

The oldest traces of hominid existence in Switzerland date back about 150,000 years. The oldest known farming settlements in Switzerland, which were found at Gächlingen, have been dated to around 5300 BC.

Founded in 44 BC, Augusta Raurica was the first Roman settlement on the Rhine and is nowadays the most important archaeological site in Switzerland

The earliest known cultural tribes of the area were members of the Hallstatt and La Tene cultures, named after the archaeological site of La Tene on the north side of Lake Neuchatel. La Tene culture developed and flourished during the late Iron Age from around 450 BC, possibly under some influence from the Greek and Etruscan civilisations. One of the most important tribal groups in the Swiss region was the Helvetii. In 58 BC, at the Battle of Bibracte, Julius Caesar's armies defeated the Helvetii. In 15 BC, Tiberius, who was destined to be the second Roman emperor and his brother, Drusus, conquered the Alps, integrating them into the Roman Empire. The area occupied by the Helvetii—the namesakes of the later Confoederatio Helvetica—first became part of Rome's Gallia Belgica province and then of its Germania Superior province, while the eastern portion of modern Switzerland was integrated into the Roman province of Raetia.

In the Early Middle Ages, from the 4th century, the western extent of modern-day Switzerland was part of the territory of the Kings of the Burgundians. The Alemanni settled the Swiss plateau in the 5th century and the valleys of the Alps in the 8th century, forming Alemannia. Modern-day Switzerland was therefore then divided between the kingdoms of Alemannia and Burgundy. The entire region became part of the expanding Frankish Empire in the 6th century, following Clovis I's victory over the Alemanni at Tolbiac in 504 AD, and later Frankish domination of the Burgundians.

Throughout the rest of the 6th, 7th and 8th centuries the Swiss regions continued under Frankish hegemony (Merovingian and Carolingian dynasties). But after its extension under Charlemagne, the Frankish empire was divided by the Treaty of Verdun in 843. The territories of nowadays Switzerland became divided into Middle Francia and East Francia until they were reunified under the Holy Roman Empire around 1000 AD.

By 1200, the Swiss plateau comprised the dominions of the houses of Savoy, Zahringer, Habsburg and Kyburg. Some regions (Uri, Schwyz, Unterwalden, later known as Waldstatten) were accorded the Imperial immediacy to grant the empire direct control over the mountain passes. When the Kyburg dynasty fell in 1264 AD, the Habsburgs under King Rudolph I (Holy Roman Emperor in 1273) extended their territory to the eastern Swiss plateau.

Old Swiss Confederacy

The Old Swiss Confederacy was an alliance among the valley communities of the central Alps. The Confederacy facilitated management of common interests (free trade) and ensured peace on the important mountain trade routes. The Federal Charter of 1291 agreed between the rural communes of Uri. Schwyz, and Nidwalden is considered the confederacy's founding document, even though similar alliances are likely to have existed decades earlier.

The Old Swiss Confederacy from 1291 (dark grey) to the sixteenth century (light grey) and its associates (blue).

By 1353 the three original cantons had joined with the cantons of Glarus and Zug and the Lucerne, Zurich and Bern city states to form the "Old Confederacy" of eight states that existed until the end of the 15th century. The expansion led to increased power and wealth for the federation. By 1460, the confederates controlled most of the territory south and west of the Rhine to the Alps and the Jura mountains, particularly after victories against the Habsburgs (Battle of Sempach, Battle of Nafels), over Charles the Bold of Burgundy during the 1470s, and the success of the Swiss mercenaries. The Swiss victory in the Swabian War against the Swabian League of Emperor Maximilian I in 1499 amounted to de facto independence within the Holy Roman Empire.

The Old Swiss Confederacy had acquired a reputation of invincibility during these earlier wars, but expansion of the federation suffered a setback in 1515 with the Swiss defeat in the Battle of Marignano. This ended the so-called "heroic" epoch of Swiss history. The success of Zwingli's Reformation in some cantons led to inter-cantonal religious conflicts in 1529 and 1531 (Wars of Kappel). It was not until more than one hundred years after these internal wars that, in 1648, under the Treaty of Westphalia, European countries recognised Switzerland's independence from the Holy Roman Empire and its neutrality.

During the Early Modern period of Swiss history, the growing authoritarianism of the patriciate families combined with a financial crisis in the wake of the Thirty Years' War led to the Swiss peasant war of 1653. In the background to this struggle, the conflict between Catholic and Protestant cantons persisted, erupting in further violence at the Battles of Villmergen in 1656 and 1712.

Napoleonic Era

The Act of Mediation was Napoleon's attempt at a compromise between the Ancient Regime and a Republic.

In 1798, the revolutionary French government conquered Switzerland and imposed a new unified constitution. This centralised the government of the country and effectively abolished the cantons and Mulhausen and Valtellina valley separated from Switzerland. The new regime, known as the Helvetic Republic, was highly unpopular. It had been imposed by a foreign invading army and destroyed centuries of tradition, making Switzerland nothing more than a French satellite state. The fierce French suppression of the Nidwalden Revolt in September of 1798 was an example of the oppressive presence of the French Army and the local population's resistance to the occupation.

When war broke out between France and its rivals, Russian and Austrian forces invaded Switzerland. The Swiss refused to fight alongside the French in the name of the Helvetic Republic. In 1803 Napoleon organised a meeting of the leading Swiss politicians from both sides in Paris. The result was the Act of Mediation which largely restored Swiss autonomy and introduced a Confederation of 19 cantons. Henceforth much of Swiss politics would concern balancing the cantons' tradition of self-rule with the need for a central government.

In 1815 the Congress of Vienna fully re-established Swiss independence and the European powers agreed to permanently recognise Swiss neutrality. Swiss troops still served foreign governments until 1860 when they fought in the Siege of Gaeta. The treaty also allowed Switzerland to increase its territory, with the admission of the cantons of Valais, Neuchatel and Geneva. Switzerland's borders have not changed since.

Federal State

The first Federal Palace is in Bern (1857). One of the three cantons presiding over the Tagsatzung (former legislative and executive council), Bern was chosen as the federal capital in 1848, mainly because of its closeness to the French speaking area.

The restoration of the power to the patriciate was only temporary. After a period of unrest with repeated violent clashes such as the Zuriputsch of 1839, civil war broke out in 1847 when some of the Catholic cantons tried to set up a separate alliance (the Sonderbundskrieg). The war lasted for less than a month, causing fewer than 100 casualties, most of which were through friendly fire. However minor the Sonderbundskrieg seems to be when compared with other European riots and wars in the 19th century, it nevertheless had a major impact on both the psychology and the society of the Swiss and of Switzerland.

The war made all Swiss understand the need for unity and strength towards its European neighbours. Swiss people from all strata of society, whether Catholic, Protestant, or from the liberal or conservative current, realised that the cantons would profit more if their economic and religious interests were merged.

Thus, while the rest of Europe was plagued by revolutionary uprisings, the Swiss drew up a constitution which provided for a federal layout, much of it inspired by the American example. This constitution provided for a central authority while leaving the cantons the right to self-government on local issues. Giving credit to those who favoured the power of the cantons (the Sonderbund Kantone), the national assembly was divided between an upper house (the Swiss Council of States, 2 representatives per canton) and a lower house (the National Council of Switzerland, representatives elected from across the country). Referenda were made mandatory for any amendment of this constitution.

Inauguration in 1882 of the Gotthard Rail Tunnel connecting the southern canton of Ticino, the longest in the world at the time.

A system of single weights and measures was introduced and in 1850 the Swiss franc became the Swiss single currency. Article 11 of the constitution forbade sending troops to serve abroad, though the Swiss were still obliged to serve Francis II of the Two Sicilies with Swiss Guards present at the Siege of Gaeta in 1860, marking the end of foreign service.

An important clause of the constitution was that it could be re-written completely if this was deemed necessary, thus enabling it to evolve as a whole rather than being modified one amendment at a time.

This need soon proved itself when the rise in population and the Industrial Revolution that followed led to calls to modify the constitution accordingly. An early draft was rejected by the population in 1872 but modifications led to its acceptance in 1874. It introduced the facultative referendum for laws at the federal level. It also established federal responsibility for defence, trade, and legal matters. In 1891, the constitution was revised with unusually strong elements of direct democracy, which remain unique even today.

20th Century

Switzerland was not invaded during either of the world wars. During World War I, Switzerland was home to Vladimir Illych Ulyanov (Lenin) and he remained there until 1917. Swiss neutrality was seriously questioned by the Grimm-Hoffmann Affair in 1917, but it was short-lived. In 1920, Switzerland joined the League of Nations, which was based in Geneva, on the condition that it was exempt from any military requirements.

During World War II, detailed invasion plans were drawn up by the Germans, but Switzerland was never attacked. Switzerland was able to remain independent through a combination of military deterrence, concessions to Germany, and good fortune as larger events during the war delayed an invasion. Under General Henri Guisan, a massive mobilisation of militia forces was ordered. The Swiss military strategy was changed from one of static defence at the borders to protect the economic heartland, to one of organised long-term attrition and withdrawal to strong, well-stockpiled positions high in the Alps known as the Reduit. Switzerland was an important base for espionage by both sides in the conflict and often mediated communications between the Axis and Allied powers.

Switzerland's trade was blockaded by both the Allies and by the Axis. Economic cooperation and extension of credit to the Third Reich varied according to the perceived likelihood of invasion and the availability of other trading partners. Concessions reached a peak after a crucial rail link through Vichy France was severed in 1942, leaving Switzerland completely surrounded by the Axis. Over the course of the war, Switzerland interned over 300,000 refugees and the International Red Cross, based in Geneva, played an important part during the conflict. Strict immigration and asylum policies as well as the financial relationships with Nazi Germany raised controversy, carrying on to this day with many Swiss banks and entities refusing to surrender the assets deposited by victims of Nazi persecution

During the war, the Swiss Air Force engaged aircraft of both sides, shooting down 11 intruding Luftwaffe planes in May and June 1940, then forcing down other intruders after a change of policy following threats from Germany. The fact that the Swiss Air Force consistently beat the Luftwaffe was a recurring embarrassment for Hitler in World War Two. The Allies acknowledged this, but the Allied Air Forces also many times intruded Swiss Air Space and made raids on several cities during the War. Over 100 Allied bombers and their crews were interned during the war. During 1944-45, Allied bombers mistakenly bombed a few places in Switzerland, among which were the cities of Schaffhausen, Basel and Zurich.

Women were granted the right to vote in the first Swiss cantons in 1959, at the federal level in 1971 and, after resistance, in the last canton Appenzell Innerrhoden in 1990. After suffrage at the federal level, women quickly rose in political significance, with the first woman on the seven member Federal Council executive being Elisabeth Kopp, who served from 1984–1989, and the first female president being Ruth Dreifuss in 1999.

In 2003, by electing Christoph Blocher, the Parliament granted the Swiss People's Party a second seat in the governing cabinet, altering the coalition which had brought stability to Swiss politics since 1959.

Switzerland joined the Council of Europe in 1963. In 1979 areas from the canton of Bern attained independence from the Bernese, forming the new canton of Jura. On 18 April 1999 the Swiss population and the cantons voted in favour of a completely revised federal constitution.

In 2002 Switzerland became a full member of the United Nations, leaving the Vatican as the last widely recognised state without full UN membership. Switzerland is a founding member of the EFTA, but is not a member of the European Economic Area. An application for membership in the European Union was sent in May 1992, but not advanced since the EEA was rejected in December 1992 when Switzerland was the only country to launch a referendum on the EEA. There have since been several referenda on the EU issue; due to a mixed reaction from the population the membership application has been frozen.

Nonetheless, Swiss law is gradually being adjusted to conform with that of the EU and the government has signed a number of bilateral agreements with the European Union. Switzerland, together with Liechtenstein, has been completely surrounded by the EU since Austria's membership in 1995. On 5 June 2005, Swiss voters agreed by a 55% majority to join the Schengen treaty, a result that was regarded by EU commentators as a sign of support by Switzerland, a country that is traditionally perceived as independent and reluctant to enter supranational bodies.

Politics

The Swiss Federal Council in 2010

The Federal Constitution adopted in 1848 is the legal foundation of the modern federal state, the second oldest in the world. A new Constitution was adopted in 1999, but did not introduce notable changes to the federal structure. It outlines basic and political rights of individuals and citizen participation in public affairs, divides the powers between the Confederation and the cantons and defines federal jurisdiction and authority. There are three main governing bodies on the federal level: the bicameral parliament (legislative), the Federal Council (executive) and the Federal Court (judicial).

The Federal Palace in Bern is the name of the building in which the Federal Assembly of Switzerland (federal parliament) and the Swiss Federal Council (executive) are housed.

The Swiss Parliament consists of two houses: the Council of States which has 46 representatives (two from each canton and one from each half-canton) who are elected under a system determined by each canton, and the National Council, which consists of 200 members who are elected under a system of proportional representation, depending on the population of each canton. Members of both houses serve for 4 years. When both houses are in joint session, they are known collectively as the Federal Assembly. Through referendums, citizens may challenge any law passed by parliament and through initiatives, introduce amendments to the federal constitution, thus making Switzerland a direct democracy.

The Federal Council constitutes the federal government, directs the federal administration and serves as collective Head of State. It is a collegial body of seven members, elected for a four-year mandate by the Federal Assembly which also exercises oversight over the Council. The President of the Confederation is elected by the Assembly from among the seven members, traditionally in rotation and for a one-year term; the President chairs the government and assumes representative functions. However, the president is a primus inter pares with no additional powers, and remains the head of a department within the administration.

The Swiss government has been a coalition of the four major political parties since 1959, each party having a number of seats that roughly reflects its share of electorate and representation in the federal parliament. The classic distribution of 2 CVP/PDC, 2 SPS/PSS, 2 FDP/PRD and 1 SVP/UDC as it stood from 1959 to 2003 was known as the "magic formula". In the 2007 Federal Council elections the seven seats in the Federal Council were distributed as follows:

- 2 Social Democrats (SPS/PSS),
- 2 Liberal Democrats (FDP/PRD),
- 2 Swiss People's Party (SVP/UDC).

Christian Democrats (CVP/PDC)

The function of the Federal Supreme Court is to hear appeals against rulings of cantonal or federal courts. The judges are elected by the Federal Assembly for six-year terms.

Direct Democracy

The Landsgemeinde is an old form of direct democracy. It is still practised in two cantons.

Swiss citizens are subject to three legal jurisdictions: the commune, canton and federal levels. The 1848 federal constitution defines a system of direct democracy (sometimes called half-direct or representative direct democracy since it is aided by the more commonplace institutions of a parliamentary democracy). The instruments of Swiss direct democracy at the federal level, known as civic rights (Volksrechte, droits civiques), include the right to submit a constitutional initiative and a referendum, both of which may overturn parliamentary decisions.

By calling a federal referendum a group of citizens may challenge a law that has been passed by Parliament, if they can gather 50,000 signatures against the law within 100 days. If so, a national vote is scheduled where voters decide by a simple majority whether to accept or reject the law. Eight cantons together can also call a referendum on a federal law.

Similarly, the federal constitutional initiative allows citizens to put a constitutional amendment to a national vote, if they can get 100,000 voters to sign the proposed amendment within 18 months. Parliament can supplement the proposed amendment with a counter-proposal, with voters having to indicate a preference on the ballot in case both proposals are accepted. Constitutional amendments, whether introduced by initiative or in Parliament, must be accepted by a double majority of both the national popular vote and a majority of the cantonal popular votes.

Cantons

The Swiss Confederation consists of 26 cantons:

The cantons have a permanent constitutional status and, in comparison with the situation in other countries, a high degree of independence. Under the Federal Constitution, all 26 cantons are equal in status. Each canton has its own constitution, and its own parliament, government and courts. However, there are considerable differences between the individual cantons, most particularly in terms of population and geographical area. Their populations vary between 15,000 (Appenzell Innerrhoden) and 1,253,500 (Zurich), and their area between 37 km2 (14 sq mi) (Basel-Stadt) and 7,105 km2 (2,743 sq mi) (Graubünden). The Cantons comprise a total of 2,889 municipalities. Within Switzerland there are two enclaves: Busingen belongs to Germany, Campione d'Italia belongs to Italy.

In a referendum held in the Austrian state of Vorarlberg on 11 May 1919 over 80% of those voting supported a proposal that the state should join the Swiss Confederation. However, this was prevented by the opposition of the Austrian Government, the Allies, Swiss liberals and non German-speaking Swiss.

Foreign Relations and International Institutions

Traditionally, Switzerland avoids alliances that might entail military, political, or direct economic action and had been neutral since the end of its expansion in 1515. Its policy of neutrality has been internationally recognised at the Congress of Vienna in 1815. Only in 2002 did Switzerland become a full member of the United Nations but it was the first state to join it by referendum. Switzerland maintains diplomatic relations with almost all countries and historically has served as an intermediary between other states. Switzerland is not a member of the European Union; the Swiss people have consistently rejected membership since the early 1990s.

The reversed Swiss flag became the symbol of the Red Cross Movement, founded in 1863 by Henri Dunant

An unusual number of international institutions have their seats in Switzerland, in part because of its policy of neutrality. Geneva is the birth place of the Red Cross and Red Crescent Movement and hosts the United Nations Human Rights Council. The European Broadcasting Union has the official headquarters in the city. Even though Switzerland is one of the most recent countries to have joined the United Nations, the Palace of Nations in Geneva is the second biggest centre for the United Nations after New York, and Switzerland was a founding member of the League of Nations.

Apart from the United Nations headquarters, the Swiss Confederation is host to many UN agencies, like the World Health Organization (WHO), the International Telecommunication Union (ITU) and about 200 other international organisations. The World Economic Forum foundation is based in Geneva. It is best known for its annual meeting in Davos which brings together top international business and political leaders to discuss important issues facing the world, including health and the environment.

Furthermore, many sport federations and organisations are located throughout the country, such as the International Ice Hockey Federation. The most important ones are probably the International Olympic Committee, in Lausanne, the FIFA (International Federation of Association Football), in Zurich, and the UEFA (Union of European Football Association), in Nyon.

Swiss Armed Forces

The Swiss Armed Forces, including the Land Forces and the Air Force, are composed of conscripts: professional soldiers constitute only about 5 percent of the military personnel, and all the rest are conscript citizens aged from 20 to 34 (in special cases up to 50) years. Being a landlocked country, Switzerland has no navy, however on lakes bordering neighbouring countries armed military patrol boats are used. Swiss citizens are prohibited from serving in foreign armies, with the exception of the Swiss Guards of the Vatican.

The structure of the Swiss militia system stipulates that the soldiers keep their own personal equipment, including all personal weapons, at home. Some organisations and political parties find this practice controversial and dangerous. Compulsory military service concerns all male Swiss citizens; women can serve voluntarily. Men usually receive military conscription orders for training at the age of 19. About two thirds of the young Swiss are found suited for service; for those found unsuited, an alternative service exists. Annually, approximately 20,000 persons are trained in recruit centres for a duration from 18 to 21 weeks. The reform "Army XXI" was adopted by popular vote in 2003, it replaced the previous model "Army 95", reducing the effectives from 400,000 to about 200,000. Of those, 120,000 are active and 80,000 are reserve units.

MOWAG Eagle Armoured Vehicles in a Military Parade

Overall, three general mobilisations have been declared to ensure the integrity and neutrality of Switzerland. The first one was held on the occasion of the Franco-Prussian War of 1870-71. The second one was decided in response to the outbreak of the First World War in August 1914. The third mobilisation of the army took place on September 1939 in response to the German attack on Poland; Henri Guisan was elected as the General-in-Chief.

Because of the neutrality policy, the army does not take part in armed conflicts in other countries, but is part of some peacekeeping missions around the world. Since 2000 the armed forces department has also maintained the Onyx intelligence gathering system to monitor satellite communications.

Following the end of the Cold War there have been a number of attempts to curb military activity or even abolish the armed forces altogether. A notable referendum on the subject, launched by an anti-militarist group, was held on the 26 November 1989. It was defeated with about two thirds of the voters against the proposal. A similar referendum, called for before, but held shortly after, the September 11 attacks in the US, was defeated by over 78% of voters.

Geography and Climate

Extending across the north and south side of the Alps, Switzerland encompasses a great diversity of landscapes and climates on a limited area of 41,285 square kilometres (15,940 sq mi). The population is about 7.8 million, resulting in an average population density of around 190 people per square kilometre (485/sq mi). The more mountainous southern half of the country is far more sparsely populated than the northern half.

Switzerland contains three basic topographical areas: the Swiss Alps on the south, the Central Plateau or middleland, and the Jura mountains on the north. The Alps are a high mountain range running across the central-south of the country, comprising about 60% of the country's total area. Among the high valleys of the Swiss Alps countless glaciers are found, totalling an area of 1,063 square kilometres. From these originate the headwaters of several major rivers, such as the Rhine, Inn, Ticino and Rhone, which flow in the four cardinal directions into the whole of Europe. The hydrographic network includes several of the largest bodies of freshwater in western Europe, among which are included Lake Geneva, Lake Constance and Lake Maggiore. Lakes and glaciers cover about

6% of the national territory. About a hundred of Switzerland's mountain peaks are close to or higher than 4,000 metres. At 4,634 m/15,203 ft, the Dufourspitze is the highest, although the Matterhorn (4,478 m/14,692 ft) is probably the most famous. Both are located within the Pennine Alps in the canton of Valais. The section of the Bernese Alps above the deep glacial Lauterbrunnen valley, containing 72 waterfalls, is well known for the Jungfrau (4,158 m/13,642 ft) and Eiger, and the many picturesque valleys in the region. In the southeast the long Engadin Valley, encompassing the St. Moritz area in canton Graubünden, is also well known; the highest peak in the neighbouring Bernina Alps is Piz Bernina (4,049 m/13,284 ft).

The more populous northern part of the country, comprising about 30% of the country's total area, is called the Middle Land. It has greater open and hilly landscapes, partly forested, partly open pastures, usually with grazing herds, or vegetables and fruit fields, but it is still hilly. There are large lakes found here and the biggest Swiss cities are in this area of the country. The largest lake is Lake Geneva (also called Lac Léman in French), in western Switzerland. The Rhone River is both the main input and output of Lake Geneva.

The Swiss climate is generally temperate, but can vary greatly between the localities, from glacial conditions on the mountaintops to the often pleasant near Mediterranean climate at Switzerland's southern tip. Summers tend to be warm and humid at times with periodic rainfall so they are ideal for pastures and grazing. The winters in the mountains alternate with sun and snow, while the lower lands tend to be more cloudy and foggy in winter. A weather phenomenon known as the fohn can occur at all times of the year, even in winter, and is characterised by a relatively warm wind, bringing air of very low relative humidity. It blows mostly on the northern side of the Alps where it can trigger dangerous avalanches.

The driest conditions persist in the southern valleys of the Valais above which valuable saffron is harvested and many wine grapes are grown, Graubünden also tends to be drier in climate and slightly colder, yet with plentiful snow in winter. The wettest conditions persist in the high Alps and in the Ticino canton which has much sun yet heavy bursts of rain from time to time. The eastern part tends to be colder than western Switzerland, yet anywhere up high in the mountains can experience a cold spell at any time of the year. Precipitation tends to be spread moderately throughout the year, with minor variations across the seasons depending on locale. Autumn is frequently the driest season, yet the weather patterns in Switzerland can be highly variable from year to year, and difficult to predict.

Switzerland's ecosystems can be particularly fragile, because of the many delicate valleys separated by high mountains, often forming unique ecologies. The mountainous regions themselves are also vulnerable, with a rich range of plants not found at other altitudes, and experience some pressure from visitors and grazing. The climatic, geological and topographical conditions of the alpine region make for a very fragile ecosystem that is particularly sensitive to climate change.

Economy

Switzerland has a stable, modern and one of the most capitalist economies in the world. It has the 2nd highest European rating in the Index of Economic Freedom 2010, while also providing large coverage through public services. The nominal per capita GDP is higher than those of the larger western European economies and Japan, ranking 6th behind Luxembourg, Norway, Qatar, Iceland and Ireland. The Swiss franc remains one of the world's strongest currencies with the lowest inflation rate.

The Greater Zurich Area, home to 1.5 million inhabitants and 150,000 companies, has taken top position in some life quality surveys.

The Engadin Valley. Tourism constitutes an important revenue for the less industrialised alpine regions.

If adjusted for purchasing power parity, Switzerland ranks 15th in the world for GDP per capita. The World Economic Forum's Global Competitiveness Report currently ranks Switzerland's economy as the most competitive in the world. For much of the 20th century, Switzerland was the wealthiest country in Europe by a considerable margin. In 2005 the median household income in Switzerland was an estimated 95,000 CHF, the equivalent of roughly 90,000 USD (as of December 2009) in nominal terms.

Switzerland is home to several large multinational corporations. The largest Swiss companies by revenue are Glencore, Nestle, Novartis, Hoffmann-La Roche, ABB and Adecco. Also notable are UBS AG, Zurich Financial Services, Credit Suisse, Swiss Re, and The Swatch Group. Switzerland is ranked as having one of the most powerful economies in the world.

Chemicals, health and pharmaceutical, measuring instruments, musical instruments, real estate, banking and insurance, tourism, and international organisations are important industries in Switzerland. The largest exported goods are chemicals (34% of exported goods), machines/electronics (20.9%), and precision instruments/watches (16.9%). Exported services amount to a third of exported goods.

Around 3.8 million people work in Switzerland. Switzerland has a more flexible job market than neighboring countries and the unemployment rate is very low. Unemployment rate increased from a low of 1.7% in June 2000 to a peak of 4.4%, as of December 2009. Population growth from net immigration is quite high, at 0.52% of population in 2004. Foreign citizen population is 21.8% as of 2004, about the same as in Australia. GDP per hour worked is the world's 17th highest, at 27.44 international dollars in 2006.

Switzerland has an overwhelmingly private sector economy and low tax rates by Western standards; overall taxation is one of the smallest of developed countries. Switzerland is an easy place to do business; Switzerland ranks 21st of 178 countries in the Ease of Doing Business Index. The slow growth Switzerland experienced in the 1990s and the early 2000s has brought greater support for economic reforms and harmonisation with the European Union. According to Credit Suisse, only about 37% of residents own their own homes, one of the lowest rates of home ownership in Europe. Housing and food price levels were 171% and 145% of the EU-25 index in 2007, compared to 113% and 104% in Germany.

Agricultural protectionism—a rare exception to Switzerland's free trade policies—has contributed to high food prices. Product market liberalisation is lagging behind many EU countries according to the OECD. Nevertheless, domestic purchasing power is one of the best in the world. Apart from agriculture, economic and trade barriers between the European Union and Switzerland are minimal and Switzerland has free trade agreements worldwide. Switzerland is a member of the European Free Trade Association (EFTA).

Education, Science and Technology

Some of the Swiss scientists who played a key role in their discipline (clockwise):

1. Leonhard Euler (maths)
2. Louis Agassiz (glaciology)
3. Auguste Piccard (aeronautics)
4. Albert Einstein (physics).

Education in Switzerland is very diverse because the constitution of Switzerland delegates the authority for the school system to the cantons. There are both public and private schools, including many private international schools. The minimum age for primary school is about six years in all cantons, but most cantons provide a free "children's school" starting at four or five years old. Primary school continues until grade four or five, depending on the school. Traditionally, the first foreign language in school was always one of the other national languages, although recently (2000) English was introduced first in a few cantons.

At the end of primary school (or at the beginning of secondary school), pupils are separated according to their capacities in several (often three) sections. The fastest learners are taught advanced classes to be prepared for further studies and the matura, while students who assimilate a little bit more slowly receive an education more adapted to their needs.

There are 12 universities in Switzerland, ten of which are maintained at cantonal level and usually offer a range of non-technical subjects. The first university in Switzerland was founded in 1460 in Basel (with a faculty of medicine) and has a tradition of chemical and medical research in Switzerland. The biggest university in Switzerland is the University of Zurich with nearly 25,000 students. The two institutes sponsored by the federal government are the ETHZ in Zurich (founded 1855) and the EPFL in Lausanne (founded 1969 as such, formerly an institute associated with the University of Lausanne) which both have an excellent international reputation.

In addition there are various Universities of Applied Sciences. In business and management studies, University of St. Gallen (HSG) and International Institute for Management Development (IMD) are the leaders. Switzerland has the second highest rate of foreign students in tertiary education, after Australia.

Many Nobel prizes were awarded to Swiss scientists, for example to the world-famous physicist Albert Einstein in the field of physics who developed his theory of relativity while working in Bern. More recently Vladimir Prelog, Heinrich Rohrer, Richard Ernst, Edmond Fischer, Rolf Zinkernagel and Kurt Wuthrich received Nobel prizes in the sciences. In total, 113 Nobel Prize winners stand in relation to Switzerland and the Nobel Peace Prize was awarded 9 times to organisations residing in Switzerland.

The LHC tunnel. The CERN is the world's largest laboratory and also the birthplace of the World Wide Web.

Geneva hosts the world's largest laboratory, the CERN, dedicated to particle physics research. Another important research center is the Paul Scherrer Institute. Notable inventions include the lysergic acid diethylamide (LSD), the scanning tunneling microscope (Nobel prize) or the very popular Velcro. Some technologies enabled the exploration of new worlds such as the pressurized balloon of Auguste Piccard and the Bathyscaphe which permitted Jacques Piccard to reach the deepest point of the world's oceans.

Switzerland Space Agency, the Swiss Space Office, has been involved in various space technologies and programs. In addition it was one of the 10 founders of the European Space Agency in 1975 and is the seventh largest contributor to the

ESA budget. In the private sector, several companies are implicated in the space industry such as Oerlikon Space or Maxon Motors who provide spacecraft structures.

Switzerland and the European Union

Switzerland voted against membership in the European Economic Area in a referendum in December 1992 and has since maintained and developed its relationships with the European Union (EU) and European countries through bilateral agreements. In March 2001, the Swiss people refused in a popular vote to start accession negotiations with the EU. In recent years, the Swiss have brought their economic practices largely into conformity with those of the EU in many ways, in an effort to enhance their international competitiveness. The economy has been growing most recently at around 3% per year. Full EU membership is a long-term objective of some in the Swiss government, but there is considerable popular sentiment against this supported by the conservative SVP party. The western French-speaking areas and the urban regions of the rest of the country tend to be more pro-EU, however with far from any significant share of the population.

The government has established an Integration Office under the Department of Foreign Affairs and the Department of Economic Affairs. To minimise the negative consequences of Switzerland's isolation from the rest of Europe, Bern and Brussels signed seven bilateral agreements to further liberalise trade ties. These agreements were signed in 1999 and took effect in 2001. This first series of bilateral agreements included the free movement of persons. A second series covering nine areas was signed in 2004 and has since been ratified. The second series includes the Schengen Treaty and the Dublin Convention. They continue to discuss further areas for cooperation.

In 2006, Switzerland approved a billion francs supportive investment in the poorer eastern European countries in support of cooperation and positive ties to the EU as a whole. A further referendum will be needed to approve 300 million francs to support Romania and Bulgaria and their recent admission. The Swiss have also been under EU and sometimes international pressure to reduce banking secrecy and to raise tax rates to parity with the EU. Preparatory discussions are being opened in four new areas: opening up the electricity market, participation in the European GNSS project Galileo, cooperating with the European centre for disease prevention and recognising certificates of origin for food products.

On 27 November 2008, the interior and justice ministers of European Union in Brussels announced Switzerland's accession to the Schengen passport-free zone from 12 December 2008. The land border checkpoints will remain in place only for goods movements, but should not run controls on people, though people entering the country had their passports checked until 29 March 2009 if they originated from a Schengen nation.

Energy, Infrastructure and Environment

Electricity generated in Switzerland is 56% from hydroelectricity and 39% from nuclear power, with 5% of the electricity generated from conventional power sources resulting in a nearly CO2-free electricity-generating network. On 18 May 2003, two anti-nuclear initiatives were turned down: Moratorium Plus, aimed at forbidding the building of new nuclear power plants (41.6% supported and 58.4% opposed), and Electricity Without Nuclear (33.7% supported and 66.3% opposed).

The former ten-year moratorium on the construction of new nuclear power plants was the result of a citizens' initiative voted on in 1990 which had passed with 54.5% Yes vs. 45.5% No votes. A new nuclear plant in the Canton of Bern is presently planned. The Swiss Federal Office of Energy (SFOE) is the office responsible for all questions relating to energy supply and energy use within the Federal Department of Environment, Transport, Energy and Communications (DETEC). The agency is supporting the 2000-watt society initiative to cut the nation's energy use by more than half by the year 2050.

Entrance of the new Lötschberg Base Tunnel, third longest railway tunnel in the world, under the old Lötschberg railway line. It is the first completed tunnel of the greater project AlpTransit.

A very dense rail network of 5,063 km (3,146 mi) carries over 350 million passengers annually. In 2007, each Swiss citizen travelled on average 2,103 km (1,307 mi) by rail, which makes them the keenest rail users. The network is administered mainly by the Federal Railways, except in Graubünden, where the 366 km (227 mi) narrow gauge railway is operated by the Rhaetian Railways and includes some World Heritage lines. The building of new railway base tunnels through the Alps is under way to reduce the time of travel between north and south.

Swiss private-public managed road network is funded by road tolls and vehicle taxes. The Swiss autobahn/autoroute system requires the purchase of a vignette (toll sticker)—which costs 40 Swiss francs—for one calendar year in order to use its roadways, for both passenger cars and trucks. The Swiss autobahn/autoroute network has a total length of 1,638 km (1,018 mi) (as of 2000) and has, by an area of 41,290 km2 (15,940 sq mi), also the one of the highest motorway densities in the world. Zurich Airport is Switzerland's largest international flight gateway, which handled 20.7 million passengers in 2007.

Switzerland has one of the best environmental records among nations in the developed world; it was one of the countries to sign the Kyoto Protocol in 1998 and ratified it in 2003. With Mexico and the Republic of Korea it forms the Environmental Integrity Group (EIG). The country is heavily active in recycling and anti-littering regulations and is one of the top recyclers in the world, with 66% to 96% of recyclable materials being recycled, depending on the area of the country.

In many places in Switzerland, household garbage disposal is charged for. Garbage (except dangerous items, batteries etc.) is only collected if it is in bags which either have a payment sticker attached, or in official bags with the surcharge paid at the time of purchase. This gives a financial incentive to recycle as much as possible, since recycling is free. Swiss health officials and police often open up garbage for which the disposal charge has not been paid and search for evidence such as old bills which connect the bag to the household/person they originated from. Fines for not paying the disposal fee range from CHF 200–500.

Demographics

Switzerland lies at the crossroads of several major European cultures that have heavily influenced the country's languages and culture. Switzerland has four official languages: German (63.7% total population share, with foreign residents; 72.5% of residents with Swiss citizenship, in 2000) in the north, east and center of the country; French (20.4%; 21.0%) to the west; Italian (6.5%; 4.3%) in the south. Romansh, a Romance language spoken locally by a small minority (0.5%; 0.6%) in the southeastern canton of Graubünden, is designated by the Federal Constitution as a national language along with German, French and Italian (Article 4 of the Constitution), and as official language if the authorities communicate with persons of Romansh language (Article 70), but federal laws and other official acts do not need to be decreed in this language. The federal government is obliged to communicate in the official languages, and in the federal parliament simultaneous translation is provided from and into German, French and Italian.

The German spoken in Switzerland is predominantly a group of Alemannic dialects collectively known as Swiss German, but written communication typically use Swiss Standard German, whilst the majority of radio and TV broadcast is (nowadays) in Swiss German as well. Similarly, there are some dialects of Franco-Provincial in rural communities in the French speaking part, known as "Suisse romande", called Vaudois, Gruerien, Jurassien, Empro, Fribourgeois, Neuchatelois, and in the Italian speaking area, Ticinese (a dialect of Lombard). Also the official languages (German, French and Italian) borrow some terms not understood outside of Switzerland, i.e. terms from other languages (German Billette from French), from similar term in another language (Italian azione used not as act but as discount from German Aktion). Learning one of the other national languages at school is obligatory for all Swiss, so many Swiss are supposed to be at least bilingual, especially those belonging to minorities.

Resident foreigners and temporary foreign workers make up about 22% of the population. Most of these (60%) are from European Union or EFTA countries. Italians are the largest single group of foreigners with 17.3% of total foreign population.

They are followed by Germans (13.2%), immigrants from Serbia and Montenegro (11.5%) and Portugal (11.3%). Immigrants from Sri Lanka, most of them former Tamil refugees, are the largest group among people of Asian origin. In the 2000s, domestic and international institutions have expressed concern about what they perceive as an increase of xenophobia, particularly in some political campaignings. However, the high proportion of foreign citizens in the country, as well as the generally unproblematic integration of foreigners, underlines Switzerland's openness.

Health

Swiss citizens are covered by a compulsory universal health-insurance, permitting access to a broad range of modern medical services. The healthcare system compares well with other European countries and patients are largely satisfied with it. In 2006 life expectancy at birth was 79 years for men and 84 years for women. It is among the highest in the world. However, spending on health is particularly high, with 11.5% of GDP (2003) and, from 1990, a steady increase is observed, reflecting the high prices of the services provided With aging populations and new healthcare technologies, health spending will likely continue to rise.

Urbanisation

Between two thirds and three quarters of the population live in urban areas. Switzerland has gone from a largely rural country to an urban one in just 70 years. Since 1935 urban development has claimed as much of the Swiss landscape as it did during the previous 2,000 years. This urban sprawl does not only affect the plateau but also the Jura and the Alpine foothills and there are growing concerns about land use. However from the beginning of the 21st century, the population growth in urban areas is higher than in the countryside.

Switzerland has a dense network of cities, where large, medium and small cities are complementary. The plateau is very densely populated with about 450 people per km^2 and the landscape continually shows signs of man's presence. The weight of the largest metropolitan areas, which are Zurich, Geneva-

Lausanne, Basel and Bern tend to increase. In international comparison the importance of these urban areas is stronger than their number of inhabitants suggests. In addition the two main centres of Zurich and Geneva are recognized for their particular great quality of life.

Religion

Switzerland has no official state religion, though most of the cantons (except Geneva and Neuchatel) recognize official churches, which are either the Catholic Church or the Swiss Reformed Church. These churches, and in some cantons also the Old Catholic Church and Jewish congregations, are financed by official taxation of adherents.

Christianity is the predominant religion of Switzerland, divided between the Catholic Church (41.8% of the population) and various Protestant denominations (35.3%). Immigration has brought Islam (4.3%, predominantly Kosovars, Bosniaks and Turks) and Eastern Orthodoxy (1.8%) as sizeable minority religions.

In a 2009 referendum, Swiss voters banned the construction of new minarets. The 2005 Eurobarometer poll found 48% to be theist, 39% expressing belief in "a spirit or life force", 9% atheist and 4% agnostic. Greeley (2003) found that 27% of the population does not believe in God.

The country is historically about evenly balanced between Catholic and Protestant, with a complex patchwork of majorities over most of the country. One canton, Appenzell, was officially divided into Catholic and Protestant sections in 1597. The larger cities (Bern, Geneva, Zurich and Basel) are predominantly Protestant. Central Switzerland, as well as Ticino, is traditionally Catholic.

The Swiss Constitution of 1848, under the recent impression of the clashes of Catholic vs. Protestant cantons that culminated in the Sonderbundskrieg, consciously defines a consociational state, allowing the peaceful co-existence of Catholics and Protestants. A 1980 initiative calling for the complete separation of church and state was resoundingly rejected, with only 21.1% voting in support.

Culture

Alphorn Concert in Vals

Switzerland is in the unusual situation of being the home of three of Europe's major languages. Swiss culture is characterised by diversity, which is reflected in a wide range of traditional customs. A region may be in some ways strongly culturally connected to the neighbouring country that shares its language, the country itself being rooted in western European culture. The linguistically isolated Romansh culture in eastern Switzerland constitutes an exception, it survives only in the upper valleys of the Rhine and the Inn and strives to maintain its rare linguistic tradition. Switzerland is home to many notable contributors to literature, art, architecture, music and sciences. In addition the country attracted a number of creative persons during time of unrest or war in Europe. Some 1000 museums are distributed through the country; the number has more than tripled since 1950. Among the most important cultural performances held annually are the Locarno International Film Festival and the Montreux Jazz Festival.

Alpine symbolism has played an essential role in shaping the history of the country and the Swiss national identity. Nowadays many mountain areas have a strong highly energetic ski resort culture in winter, and a hiking (wandering) culture in summer. Some areas throughout the year have a recreational culture that caters to tourism, yet the quieter seasons are spring and autumn when there are fewer visitors and a higher ratio of Swiss. A traditional farmer and herder culture also predominates in many areas and small farms are omnipresent outside the cities. Folk art is kept alive in organisations all over the country. In Switzerland it is mostly expressed in music, dance, poetry, wood carving and embroidery. The alphorn, a trumpet-like musical instrument made of wood, has become alongside yodelling and the accordion an epitome of traditional Swiss music.

Literature

As the Confederation, from its foundation in 1291, was almost exclusively composed of German-speaking regions, the earliest forms of literature are in German. In the 18th century French became the fashionable language in Bern and elsewhere, while the influence of the French-speaking allies and subject lands was more marked than before.

Among the classics of Swiss German literature are Jeremias Gotthelf (1797–1854) and Gottfried Keller (1819–1890). The undisputed giants of 20th century Swiss literature are Max Frisch (1911–91) and Friedrich Durrenmatt (1921–90), whose repertoire includes Die Physiker (The Physicists) and Das Versprechen (The Pledge), released in 2001 as a Hollywood film.

Prominent French-speaking writers were Jean-Jacques Rousseau (1712–1778) and Germaine de Staël (1766–1817). More recent authors include Charles Ferdinand Ramuz (1878–1947), whose novels describe the lives of peasants and mountain dwellers, set in a harsh environment and Blaise Cendrars (born Frederic Sauser, 1887–1961). Also Italian and Romansh-speaking authors contributed but in more modest way given their small number.

The probably most famous Swiss literary creation, Heidi, the story of an orphan girl who lives with her grandfather in the Alps, was one of the most popular children's books ever and has come to be a symbol of Switzerland. Her creator, Johanna Spyri (1827–1901), wrote a number of other books around similar themes.

Media

The freedom of the press and the right to free expression is guaranteed in the federal constitution of Switzerland. The Swiss News Agency (SNA) broadcasts information around-the-clock in three of the four national languages—on politics, economics, society and culture. The SNA supplies almost all Swiss media and a couple dozen foreign media services with its news.

Switzerland has historically boasted the greatest number of newspaper titles published in proportion to its population and size. The most influential newspapers are the German-language Tages-Anzeiger and Neue Zürcher Zeitung NZZ, and the French-language Le Temps, but almost every city has at least one local newspaper. The cultural diversity accounts for a large number of newspapers.

In contrast to the print media, the broadcast media has always been under greater control of the government. The Swiss Broadcasting Corporation, whose name was recently changed to SRG SSR idee suisse, is charged with the production and broadcast of radio and television programs. SRG SSR studios are distributed throughout the various language regions. Radio content is produced in six central and four regional studios while the television programs are produced in Geneva, Zurich and Lugano. An extensive cable network also allows most Swiss to access the programs from neighboring countries.

Sports

Skiing, snowboarding and mountaineering are among the most popular sports in Switzerland, the nature of the country being particularly suited for such activities. Winter sports are practiced by the natives and tourists since the second half of the 19th century with the invention of bobsleigh in St. Moritz. The first world ski championships were held in Murren (1931) and St. Moritz (1934). The latter town hosted the second Winter Olympic Games in 1928 and the fifth edition in 1948. Among the most successful skiers and world champions are Pirmin Zurbriggen and Didier Cuche.

Food

The cuisine of Switzerland is multi-faceted. While some dishes such as fondue, raclette or rösti are omnipresent through the country, each region developed its own gastronomy according to the differences of climate and languages. Traditional Swiss cuisine uses ingredients similar to those in other European countries, as well as unique dairy products and cheeses such as Gruyere or Emmental, produced in the valleys of Gruyeres and Emmental. The number of fine-dining establishments is high, particularly in western Switzerland.

Chocolate had been made in Switzerland since the 18th century but it gained its reputation at the end of the 19th century with the invention of modern techniques such as conching and tempering which enabled its production on a high quality level. Also a breakthrough was the invention of milk chocolate in 1875 by Daniel Peter. The Swiss are the world's largest consumers of chocolate.

The most popular alcoholic drink in Switzerland is wine. Switzerland is notable for the variety of grapes grown because of the large variations in terroirs, with their specific mixes of soil, air, altitude and light. Swiss wine is produced mainly in Valais, Vaud (Lavaux), Geneva and Ticino, with a small majority of white wines. Vineyards have been cultivated in Switzerland since the Roman era, even though certain traces can be found of a more ancient origin. The most widespread varieties are the Chasselas (called Fendant in Valais) and Pinot Noir. The Merlot is the main variety produced in Ticino.

2

Italy: Once a Great Power

Italy; Italian: Italia, officially the Italian Republic (Italian: Repubblica italiana), is a country located partly on the European Continent and partly on the Italian Peninsula in Southern Europe and on the two largest islands in the Mediterranean Sea, Sicily and Sardinia. Italy shares its northern, Alpine boundary with France, Switzerland, Austria and Slovenia. The independent states of San Marino and the Vatican City are enclaves within the Italian Peninsula, and Campione d'Italia is an Italian exclave in Switzerland. The territory of Italy covers 301,338 km^2 and is influenced by a temperate seasonal climate. With 60.2 million inhabitants, it is the sixth most populous country in Europe, and the twenty-third most populous in the world.

The land known as Italy today has been the cradle of European cultures and peoples, such as the Etruscans and the Romans. Italy's capital, Rome, was for centuries the political centre of Western civilisation, as the capital of the Roman Empire. After its decline, Italy would endure numerous invasions by foreign peoples, from Germanic tribes such as the Lombards and Ostrogoths, to the Normans and later, the Byzantines, among others. Centuries later, Italy would become the birthplace of the Renaissance, an immensely fruitful intellectual movement that would prove to be integral in shaping the subsequent course of European thought.

Through much of its post-Roman history, Italy was fragmented into numerous kingdoms and city-states (such as the Kingdom of Sardinia, the Kingdom of the Two Sicilies and the Duchy of Milan), but was unified in 1861, a tumultuous period in history known as the "Risorgimento". In the late 19th century, through World War I, and to World War II, Italy possessed a colonial empire, which extended its rule to Libya, Eritrea, Italian Somaliland, Ethiopia, Albania, Rhodes, the Dodecanese and a concession in Tianjin, China.

Modern Italy is a democratic republic. It has been ranked the world's eighteenth most-developed country and its Quality-of-Life Index has been ranked in the top ten in the world. Italy enjoys a very high standard of living, and has a high nominal GDP per capita. It is a founding member of what is now the European Union and North Atlantic Treaty Organization. Italy is also a member of the G8 and G20. It has the world's seventh-largest nominal GDP, tenth highest GDP (PPP) and the fifth highest government budget in the world. It is also a member state of the Organisation for Economic Co-operation and Development, the World Trade Organization, the Council of Europe, and the Western European Union. Italy has the world's ninth-largest defence budget and shares NATO's nuclear weapons.

Italy plays a prominent role in European and global military, cultural and diplomatic affairs, and it is affiliated with worldwide organizations such as the Food and Agriculture Organization, World Food Programme, International Fund for Agricultural Development, Global Forum, and the NATO Defence College, which are headquartered in Rome. The country's European political, social and economic influence make it a major regional power, alongside the United Kingdom, France, Germany, and Russia and Italy has been classified in a study, measuring hard power, as being the eleventh greatest worldwide national power. The country has a high public education level, high labour force, is a globalised nation, and also has 2009's sixth best international reputation. Italy also has the world's nineteenth highest life expectancy, and in 2000 its healthcare system was ranked the second best in the world by the World Health Organization Report In 2007 it was the world's fifth most visited country, with over 43.7 million international arrivals, and boasts a long tradition in the arts, science and technology, including the world's highest number of UNESCO World Heritage Sites to date.

Etymology

The origin of the term Italia, from Latin: Italia, is uncertain. According to one of the more common explanations, the term was borrowed through Greek from the Oscan Viteliu, meaning "land of young cattle". The bull was a symbol of the southern Italian tribes and was often depicted goring the Roman wolf as a defiant symbol of free Italy during the Samnite Wars.

The name Italia originally applied only to a part of what is now Southern Italy—according to Antiochus of Syracuse, the southern portion of the Bruttium peninsula (modern Calabria). But by his time Oenotria and Italy had become synonymous, and the name also applied to most of Lucania as well. The Greeks gradually came to apply the name "Italia" to a larger region, but it was not until the time of the Roman conquests that the term was expanded to cover the entire peninsula.

History

Excavations throughout Italy reveal a modern human presence dating back to the Palaeolithic period, some 200,000 years ago. In the 8th and 7th centuries BC Greek colonies were established all along the coast of Sicily and the southern part of the Italian Peninsula became known as Magna Graecia.

Ancient Rome was at first a small agricultural community founded circa the 8th century BC that grew over the course of the centuries into a colossal empire encompassing the whole Mediterranean Sea, in which Ancient Greek and Roman cultures merged into one civilization. This civilization was so influential that parts of it survive in modern law, administration, philosophy and arts, forming the ground that Western civilization is based upon.

In its twelve-century existence it transformed itself from monarchy to republic and finally to autocracy. In steady decline since the 2nd century AD, the empire finally broke into two parts in 285 AD: the Western Roman Empire and the Byzantine Empire in the East. The western part under the pressure of

Goths finally dissolved, leaving the Italian peninsula divided into small independent kingdoms and feuding city states for the next 14 centuries, and leaving the eastern part sole heir to the Roman legacy.

Middle Ages

In the sixth century AD the Byzantine Emperor Justinian reconquered Italy from the Ostrogoths. The invasion of a new wave of Germanic tribes, the Lombards, doomed his attempt to resurrect the Western Roman Empire but the repercussions of Justinian's failure resounded further still. For the next thirteen centuries, whilst new nation-states arose in the lands north of the Alps, the Italian political landscape was a patchwork of feuding city states, petty tyrannies, and foreign invaders.

For several centuries the armies and Exarchates led by the Exarchate of Ravenna, Justinian's successors, were a tenacious force in Italian affairs-strong enough to prevent other powers such as the Arabs, the Holy Roman Empire, or the Papacy from establishing a unified Italian Kingdom, but too weak to unify and control the region.

Italy was divided for centuries into small city-states.

Italy's regions were eventually subsumed by their neighbouring empires with their conflicting interests and would remain divided up to the 19th century. It was during this vacuum of authority that the region saw the rise of the Signoria and the Comune. In the anarchic conditions that often prevailed in medieval Italian city-states, people looked to strong men to restore order and disarm the feuding elites. In times of anarchy or crisis, cities sometimes offered the Signoria to individuals perceived as strong enough to save the state, most notably the Della Scala family in Verona, the Visconti in Milan and the Medici in Florence.

Italy during this period became notable for its merchant Republics. These city-states, oligarchical in reality, had a dominant merchant class which under relative freedom nurtured academic and artistic advancement. The four classic Maritime Republics in Italy were Venice, Genoa, Pisa and Amalfi. Venice and Genoa were Europe's gateways to trade with the East, with the former producer of the renowned venetian glass. Florence was the capital of silk, wool, banks and jewelry. The Maritime Republics were heavily involved in the Crusades, taking advantage of the new political and trading opportunities, most evidently in the conquest of Zara and Constantinople funded by Venice.

During the late Middle Ages Italy was divided into smaller city-states and territories: the kingdom of Naples controlled the south, the Republic of Florence and the Papal States the centre, the Genoese and the Milanese the north and west, and the Venetians the east.

Renaissance

The unique political structures of late Middle Ages Italy and its dynamic social climate and florescent trade allowed the emergence of a unique cultural efflorescence. Italy never regained the unity it once had in the days of the Roman Empire and throughout the Middle Ages was divided into smaller city states and territories: the kingdom of Naples controlled the south, the Republic of Florence and the Papal States the center, the Genoese and the Milanese the north and west, and the Venetians the east.

Fifteenth-century Italy was one of the most urbanised areas in Europe. Most historians agree that the ideas that characterized the Renaissance and their earliest apologists and supporters had their origin in late 13th century Florence or gravitated in or around Florence, as well as the other rival city-states. The Renaissance achieved its epitome, in particular with the writings of Dante Alighieri (1265–1321) and Francesco Petrarch (1304–1374), Bocaccio, as well as the paintings of great masters starting with Giotto di Bondone (1267–1337). The Renaissance was an extremely important period in Italian history, and in European history, and brought along numerous political, philosophical, literary, cultural, social and religious reforms.

The Renaissance was so called because it was a "rebirth" of many classical ideas that had long been buried in the chapters of classical Antiquity. One could argue that the fuel for this rebirth was the rediscovery of ancient texts that had been almost 'forgotten' by Western civilization, but were preserved in some monastic libraries or private libraries of powerful and wealthy patrons. Some would argue that there were translations of Greek and Arabic texts into Latin from the Islamic world that found their way into Italy and contributed to the Italian/ European Renaissance. However, most of the manuscripts were either already in the Italian Peninsula or in 'Greece' and were taken to Italy in the centuries preceding the Renaissance by the Italians themselves (by the traders who travelled regularly to the Eastern Mediterranean, including Greece) and by Byzantine Greeks who migrated to Italy during the onslaught of the Ottoman empire, against the Byzantine Empire in the 1400s, and specially after 1453, once the Ottomans had conquered the Byzantine capital, Constantinople. These Byzantines fled the Turks, sometimes carrying precious manuscripts and their knowledge (Greek and Ancient Greek) and while fixating themselves in Italy made a discreet but crucial contribution to the Renaissance.

Renaissance scholars such as Niccolo de' Niccoli and Poggio Bracciolini scoured the libraries in search of works by such classical authors as Plato, Cicero and Vitruvius. The works of ancient Greek and Hellenistic writers (such as Plato, Aristotle, Euclid, and Ptolemy) and Muslim scientists were diffused in the Christian world, providing new intellectual material for European scholars.

The Black Death pandemic in 1348 left its mark on Italy by killing one third of the population. However, the recovery from the disaster of the Black Death led to a resurgence of cities, trade and economy which greatly stimulated the successive phase of the Humanism and Renaissance (15th-16th centuries) when Italy again returned to be the center of Western civilization, strongly influencing the other European countries with Courts like Este in Ferrara and De Medici in Florence.

Florence became Italy's main centre of the Renaissance. Numerous artists, such as Michelangelo, Leonardo da Vinci and Botticelli worked in the city. Its economy flourished, and according to the Encyclopaedia Britannica, Florence from the 14th century to the 16th century was one of Europe's greatest cities, and its numerous museums, palazzi and churches, such as the Pitti Palace and the Uffizi have been described by the encyclopedia as works of art themselves.

Rome was also a city particularly affected by the Renaissance. This period of reform changed the city's face dramatically, with works like the Pieta by Michelangelo and the frescoes of the Borgia Apartment. Rome reached the highest point of splendour under Pope Julius II (1503–1513) and his successors Leo X and Clement VII, both members of the Medici family. In this twenty-years period Rome became one of the greatest centres of art in the world. The old St. Peter's Basilica built by Emperor Constantine the Great, was re-built mainly by Michelangelo, who in Rome became one the most famous painters of Italy creating frescos in the Cappella Niccolina, the Villa Farnesina, the Raphael's Rooms, plus many other famous paintings. Michelangelo started the decoration of the ceiling of the Sistine Chapel and executed the famous statue of the Moses for the tomb of Julius. Rome lost in part its religious character, becoming increasingly a true Renaissance city, with a great number of popular feasts, horse races, parties, intrigues and licentious episodes. Its economy was rich, with the presence of several Tuscan bankers, including Agostino Chigi, who was a friend of Raphael and a patron of arts. Before his early death, Raphael also promoted for the first time the preservation of the ancient ruins.

Foreign Domination and Enlightenment (16th–19th Centuries)

A map depicting Western Europe's borders after the Treaties of Utrecht and Rastatt.

During the 18th century, Italy was one of the main centres of the Grand Tour, when many foreign aristocrats came to appreciate Italian culture and art.

After a century where the fragmented system of Italian states and principalities were able to maintain a relative independence and a balance of power in the peninsula, in 1494 the French king Charles VIII opened the first of a series of invasions, lasting half of the sixteenth century, and a competition between France and Spain for the possession of the country.

Ultimately Spain prevailed (the Treaty of Cateau-Cambresis in 1559 recognised the Spanish possession of the Duchy of Milan and the Kingdom of Naples) and for almost two centuries became the hegemon in Italy.

The holy alliance between Habsburg Spain and the Holy See resulted in the systematic persecution of any Protestant movement, with the result that Italy remained a Catholic country with marginal Protestant presence. During its long rule on Italy, the Spanish Empire systematically spoiled the country and imposed heavy taxation. It interfered and held a tight grip over the affairs of the Vatican. Moreover, Spanish administration was slow and inefficient, and its social consequences in the long term, in Southern Italy, where Spanish rule was effective, have lasted till the current age.

Austria succeeded Spain as hegemon in Italy after the Peace of Utrecht (1713), having acquired the State of Milan and the Kingdom of Naples. The Austrian domination, thanks to the Enlightenment embraced by Habsburgic emperors, somewhat improved the situation. The northern part of Italy, under the direct control of Vienna, gained economic dynamism and intellectual fervour.

The main Italian cities, such as Milan, Rome, Turin, Venice, Florence and Naples became fertile grounds for intellectual discussion and thought, and several Italian philosophers and literary figures were active at the time, such as the Milanese Cesare, Marquis of Beccaria-Bonesana, better known as Cesare Beccaria, or Antonio Genovesi. Leopold I, Grand Duke of Tuscany or also known as Leopold II of the Holy Roman Empire, abolished the death penalty and torture in the Grand Duchy of Tuscany.

Italy in the 1700s was an important stop in the European Grand Tour, a period in which foreign, mostly British, aristocrats toured France, Italy and Greece to appreciate their arts and cultures, and their monuments. With the discovery of the classical ruins of Pompeii and Herculaneum in 1748, and the restoration of the derelict parts of the surviving ancient monuments in Rome, figures such as Goethe, Shelley, Keats and Byron toured the country. Cities such as Venice and Rome were the major attractions, and Naples, Florence, Turin, Sicily, and to some extent, Milan, were popular too. Keats famously said that "Italy is the paradise of exiles".

The French Revolution and the Napoleonic Wars (1796–1815) stirred the ideas of equality, democracy, law and nation which many in Italy endorsed and even supported as the basis on which they could and eventually would build national unity in Italy. This unity, or creation of modern Italy was yet to come in the second half of the nineteenth century. The plague repeatedly returned to haunt Italy throughout the 14th to 17th centuries. Italy's last major epidemic occurred in 1656 in Naples. In northern Italy, a report of 1767 noted that there had been famine in 111 of the previous 316 years and only sixteen good harvests. Italy's population between 1700 and 1800 rose by about one-third, to 18 million.

Italian Unification (1816-1861) and Kingdom of Italy (1861–1946)

Main articles: Kingdom of Italy (1861–1946), Risorgimento, Italian unification, History of Italy as a monarchy and in the World Wars, Italian Campaign (World War I), Italian fascism, Italian Social Republic, and Military history of Italy during World War II

The creation of the Kingdom of Italy was the result of efforts by Italian nationalists and monarchists loyal to the House of Savoy to establish a united kingdom encompassing the entire Italian Peninsula. In the context of the 1848 liberal revolutions that swept through Europe, an unsuccessful war was declared on Austria.

Giuseppe Garibaldi Leading the Expedition of the Thousand

Giuseppe Garibaldi, popular amongst southern Italians, led the Italian republican drive for unification in southern Italy, while the northern Italian monarchy of the Kingdom of Piedmont-Sardinia whose government was led by Camillo Benso, conte di Cavour, had the ambition of establishing a united Italian state under its rule. The kingdom successfully challenged the Austrian Empire in the Second Italian War of Independence with the help of Napoleon III, liberating the Lombardy-Venetia. It established Turin as capital of the newly formed state. In 1865 the capital was moved to Florence.

In 1866, Victor Emmanuel II aligned the kingdom with Prussia during the Austro-Prussian War, waging the Third Italian War of Independence which allowed Italy to annex Venice. In 1870, as France during the disastrous Franco-Prussian War abandoned its positions in Rome, Italy rushed to fill the power gap by taking over the Papal State from French sovereignty. Italian unification finally was achieved, and shortly afterwards Italy's capital was moved from Florence to Rome. Whilst keeping the monarchy, the government became a parliamentary system, run by liberals.

As Northern Italy became industrialized and modernized, Southern Italy and agricultural regions of the north remained under-developed and stagnant, forcing millions of people to migrate to the emerging Industrial Triangle or abroad. The Sardinian Statuto Albertino of 1848, extended to the whole Kingdom of Italy in 1861, provided for basic freedoms, but the electoral laws excluded the non-propertied and uneducated classes from voting. In 1913, male universal suffrage was adopted. The Italian Socialist Party increased in strength, challenging the traditional liberal and conservative organisations. The high point of Italian emigration was 1913, when 872,598 persons left Italy.

Starting from the last two decades of the nineteenth century, Italy developed into a colonial power by forcing Somalia, Eritrea and later Libya and the Dodecanese under its rule. During World War I, Italy at first stayed neutral but in 1915 signed the Treaty of London, entering Entente on the promise of receiving Trento, Trieste, Istria, Dalmatia and parts of Ottoman Empire. During the war, 600,000 Italians died, and the economy collapsed. Under the Peace Treaty of Saint-Germain, Italy obtained just Bolzano-Bozen, Trento, Trieste and Istria in a victory described as "mutilated" by the public.

The turbulence that followed the devastation of World War I, inspired by the Russian Revolution, led to turmoil and anarchy. The liberal establishment, fearing a socialist revolution, started to endorse the small National Fascist Party, led by Benito Mussolini. In October 1922 the fascists attempted a coup (the Marcia su Roma, "March on Rome"), but king Victor Emmanuel III ordered the army not to intervene, instead forming an alliance with Mussolini. Over the next few years, Mussolini banned all political parties and curtailed personal liberties, thus forming a dictatorship.

In 1935, Mussolini invaded Ethiopia. This resulted in international alienation and along with other factors, led to Italy's withdrawal from the League of Nations. A first pact with Nazi Germany was concluded in 1936, and a second in 1938. Italy strongly supported Franco in the Spanish civil war. The country was opposed to Adolf Hitler's annexations of Austria, but did not interfere with it. Italy supported Germany's annexation of Sudetenland, however.

On 7 April 1939, Italy occupied Albania, a de facto protectorate for decades, and entered World War II in 1940, taking part in the late stages of the Battle of France. Mussolini, wanting a quick victory like Hitler's Blitzkriegs in Poland and France, invaded Greece in October 1940 via Albania but was forced to accept a humiliating defeat after a few months. At the same time, Italy, after initially conquering British Somalia, saw an allied counter-attack lead to the loss of all possessions in the Horn of Africa. Italy was also defeated by British forces in North Africa and was only saved by the urgently dispatched German Africa Corps led by Erwin Rommel.

Italy was invaded by the Allies in June 1943, leading to the collapse of the fascist regime and the arrest of Mussolini. In September 1943, Italy surrendered. The country remained a battlefield for the rest of the war, as the allies were moving up from the south as the north was the base for loyalist Italian fascist and German Nazi forces. The whole picture became more complex by the activity of the Italian partisans; see Italian resistance movement. The Nazis left the country on 25 April 1945 and the remaining Italian fascist forces eventually disbanded. Nearly half a million Italians (including civilians) died between June 1940 and May 1945. An estimated 200,000 partisans took part in the Resistance, and German or fascist forces killed some 70,000 Italians (including both partisans and civilians) for Resistance activities. At least 54,000 Italian prisoners of war died in the Soviet Union.

Italian Republic (1946–present)

In 1946, Victor Emmanuel III's son, Umberto II, was forced to abdicate. Italy became a republic after a referendum held on 2 June 1946, a day celebrated since as Republic Day. This was also the first time in Italy that Italian women were entitled to vote. The Republican Constitution was approved and came into force on 1 January 1948. Under the Paris Peace Treaties of 1947, the eastern border area was lost to Yugoslavia, and, later, the free territory of Trieste was divided between the two states.

Fears in the Italian electorate of a possible Communist takeover proved crucial for the first universal suffrage electoral outcome on the 18th of April 1948 when the Christian Democrats, under the leadership of Alcide De Gasperi, won the election with 48 percent of the vote. In the 1950s Italy became a member of NATO and allied itself with the United States. The Marshall Plan helped revive the Italian economy which, until the 1960s, enjoyed a period of sustained economic growth commonly called the "Economic Miracle". In 1957, Italy was a founder member of the European Economic Community (EEC), which became the European Union (EU) in 1993.

From the late 1960s till late 1980s the country experienced a hard economic crisis and the Years of Lead, a period characterized by widespread social conflicts and terrorist acts carried out by extra-parliamentary movements. The Years of Lead culminated in the assassination of the Christian Democrat leader Aldo Moro in 1978, bringing to an end the "Historic Compromise" between the DC and the Communist Party. In the 1980s, for the first time since 1945, two governments were led by non-Christian-Democrat premiers: a republican (Giovanni Spadolini) and a socialist (Bettino Craxi); the Christian Democrats remained, however, the main force supporting the government.

The Socialist Party (PSI), led by Bettino Craxi, became more and more critical of the Communists and of the Soviet Union; Craxi himself pushed in favour of US president Ronald

Reagan's positioning of Pershing missiles in Italy, a move the Communists hotly contested.

Geography

Italy is located in Southern Europe and comprises the boot-shaped Italian Peninsula and a number of islands including the two largest, Sicily and Sardinia. Although the country occupies the Italian peninsula and most of the southern Alpine basin, some of Italy's territory extends beyond the Alpine basin and some islands are located outside the Eurasian continental shelf. These territories are the comuni of: Livigno, Sexten, Innichen, Toblach (in part), Chiusaforte, Tarvisio, Graun im Vinschgau (in part), which are all part of the Danube's drainage basin, while the Val di Lei constitutes part of the Rhein's basin and the island comune of Lampedusa e Linosa is on the African continental shelf. The country's total area is 301,230 km^2, of which 294,020 km^2 is land and 7,210 km^2 is water.

Including the islands, Italy has a coastline and border of 7,600 km on the Adriatic, Ionian, Tyrrhenian seas (740 km), and borders shared with France (488 km), Austria (430 km), Slovenia (232 km) and Switzerland; San Marino (39 km) and Vatican City (3.2 km), both enclaves, account for the remainder.

The Apennine Mountains form the peninsula's backbone, the Alps form its northern boundary. The Po, Italy's longest river, flows from the Alps on the western border with France and crosses the Padan plain on its way to the Adriatic Sea. The five largest lakes are, in order of diminishing size:

The Italian Alps.

* Garda (367.94 km^2/142 sq mi)
* Maggiore (212.51 km^2/82 sq mi)
* Como (145.9 km^2/56 sq mi)
* Trasimeno (124.29 km^2/48 sq mi)
* Bolsena (113.55 km^2/44 sq mi).

Volcanism

The country is situated at the meeting point of the Eurasian Plate and the African Plate, leading to considerable seismic and volcanic activity. There are 14 volcanoes in Italy, three of which are active: Etna, Stromboli and Vesuvius. Vesuvius is the only active volcano in mainland Europe and is most famous for the destruction of Pompeii and Herculanum. Several islands and hills have been created by volcanic activity, and there is still a large active caldera, the Campi Flegrei north-west of Naples.

Climate

The climate of Italy is highly diverse and can be far from the stereotypical Mediterranean climate, depending on location. Most of the inland northern regions of Italy, for example Piedmont, Lombardy and Emilia-Romagna, have a continental climate often classified as humid subtropical (Koppen climate classification Cfa).

The coastal areas of Liguria and most of the peninsula south of Florence generally fit the Mediterranean stereotype (Koppen climate classification Csa). Conditions on peninsular coastal areas can be very different from the interior's higher ground and valleys, particularly during the winter months when the higher altitudes tend to be cold, wet, and often snowy. The coastal regions have mild winters and warm and generally dry summers, although lowland valleys can be quite hot in summer.

Nature

Italy is one of the richest countries in Europe and in the Mediterranean basin in terms of species biodiversity. Traditionally it was estimated to comprise about 5,500 vascular plant species. However, as of 2004, 6,759 species are recorded in the Data bank of Italian vascular flora. (9,000 plant species if non-vascular species are included, half Europe's total). The nation has one of the highest levels of faunal biodiversity in Europe with over 57,000 species recorded (more than a third of all European fauna). This is due to Italy's Southerly geographical position, surrounded by the Mediterranean sea. There are 8,000 km of coastline and the peninsula is in the centre of the Mediterranean Sea, forming a corridor between central Europe and North Africa. Italy also receives species from the Balkans, Eurasia and the Middle East. Of animal species present, 86% of the Italian fauna is land-based, 14% is aquatic. Insects represent about two thirds of all of Italy's fauna.

Government and Politics

The politics of Italy take place in a framework of a parliamentary, democratic republic, and of a multi-party system. Executive power is exercised collectively by the Council of Ministers, which is led by a President (Presidente del Consiglio dei Ministri), informally referred to as "premier" or primo ministro (that is, "prime minister"). Legislative power is vested in the two houses of Parliament primarily, and secondarily in the Council of Ministers. The judiciary is independent of the executive and the legislative. Italy has been a democratic republic since 2 June 1946, when the monarchy was abolished by popular referendum. The constitution was promulgated on 1 January 1948.

Giorgio Napolitano is the President of the Italian Republic, whilst Silvio Berlusconi is the nation's Prime Minister (President of the Council of Ministers).

The President of the Italian Republic (Presidente della Repubblica) is elected for seven years by the parliament sitting jointly with a small number of regional delegates. As the head of state, the President of the Republic represents the unity of the nation and has many of the duties previously given to the King of Italy. The president serves as a point of connection between the three branches of power: he is elected by the lawmakers, he appoints the executive, he is the president of the judiciary and he is also the commander-in-chief of the armed forces. The Quirinal Palace, which is the official residence of the President of the Italian Republic.

The president nominates the Prime Minister, who proposes the other ministers (formally named by the president). The Council of Ministers must obtain a confidence vote from both houses of Parliament. Legislative bills may originate in either house and must be passed by a majority in both. Italy elects a parliament consisting of two houses, the Chamber of Deputies

(Camera dei Deputati), which has 630 members and the Senate of the Republic (Senato della Repubblica), comprising 315 elected members and a small number of senators for life). Legislation may originate in either house and must be passed in identical form by a majority in each. The houses of parliament are popularly and directly elected through a complex electoral system (latest amendment in 2005) which combines proportional representation with a majority prize for the largest coalition. All Italian citizens 18 years of age and older can vote. However, to vote for the Senate, the voter must be 25 or older.

The electoral system for the Senate is based upon regional representation. As of 15 May 2006 there are seven life senators (of which three are former Presidents). Both houses are elected for a maximum of five years, but both may be dissolved by the President before the expiration of their normal term if the Parliament is unable to elect a stable government. In post-war history, this has happened in 1972, 1976, 1979, 1983, 1994, 1996, and 2008.

A peculiarity of the Italian Parliament is the representation given to Italian citizens permanently living abroad (about 2.7 million people). Among the 630 Deputies and the 315 Senators there are respectively 12 and 6 elected in four distinct overseas constituencies. These members of Parliament were elected for the first time in April 2006, and they have the same powers as those of members elected in Italy.

Law

The Italian judicial system is based on Roman law modified by the Napoleonic code and later statutes. The Supreme Court of Cassation is the court of last resort for most disputes. The Constitutional Court of Italy (Corte Costituzionale) rules on the conformity of laws with the Constitution and is a post-World War II innovation.

Foreign Relations

Italy was a founding member of the European Community—now the European Union (EU). Italy was admitted to the United Nations in 1955 and is a member and strong supporter of the North Atlantic Treaty Organization (NATO), the

Organisation for Economic Co-operation and Development (OECD), the General Agreement on Tariffs and Trade/World Trade Organization (GATT/WTO), the Organization for Security and Co-operation in Europe (OSCE), the Council of Europe, and the Central European Initiative. Its recent turns in the rotating Presidency of international organisations include the Conference for Security and Co-operation in Europe (CSCE), the forerunner of the OSCE, in 1994; G8; and the EU in 2001 and from July to December 2003.

Italy supports the United Nations and its international security activities. Italy deployed troops in support of UN peacekeeping missions in Somalia, Mozambique, and East Timor and provides support for NATO and UN operations in Bosnia, Kosovo and Albania. Italy deployed over 2,000 troops to Afghanistan in support of Operation Enduring Freedom (OEF) in February 2003. Italy still supports international efforts to reconstruct and stabilize Iraq, but it has withdrawn its military contingent of some 3,200 troops as of November 2006, maintaining only humanitarian workers and other civilian personnel. In August, 2006 Italy sent about 2,450 soldiers to Lebanon for the United Nations' peacekeeping mission UNIFIL. Furthermore, since 2 February 2007 an Italian, Claudio Graziano, is the commander of the UN force in the country.

Military

The Italian armed forces are under the command of the Supreme Defence Council, presided over by the President of the Italian Republic. In 2008 the military had 186,798 personnel on active duty, along with 114,778 in the national gendarmerie. As part of NATO's nuclear sharing strategy Italy also hosts 90 United States nuclear bombs, located in the Ghedi Torre and Aviano air bases. Total military spending in 2007 was $33.1 billion, equal to 1.8% of national GDP.

The Italian armed forces are divided into four branches:

Army

The Italian Army (Esercito Italiano) is the ground defence force of the Italian Republic. It has recently become a professional all-volunteer force of active-duty personnel, numbering 109,703 in 2008. Its best-known combat vehicles are the Dardo infantry fighting vehicle, the Centauro tank destroyer and the Ariete tank, and among its aircraft the Mangusta attack helicopter, recently deployed in UN missions. The Esercito Italiano also has at its disposal a large number of Leopard 1 and M113 armored vehicles.

Navy

The Italian Navy (Marina Militare) in 2008 had a strength of 43,882 and ships of every type, such as aircraft carriers, destroyers, modern frigates, submarines, amphibious ships, and other smaller ships such as oceanographic research ships The Marina Militare is now equipping itself with a bigger aircraft carrier, (the Cavour), new destroyers, submarines and multipurpose frigates. In modern times the Italian Navy, being a member of the North Atlantic Treaty Organization (NATO), has taken part in many coalition peacekeeping operations around the world.

Air Force

The Italian Air Force in 2008 has a strength of 43,882 and operates 585 aircraft, including 219 combat jets and 114 helicopters. As a stopgap and as replacement for leased Tornado ADV interceptors, the AMI has leased 30 F-16A Block 15 ADF and four F-16B Block 10 Fighting Falcons, with an option for more.

The coming years also will see the introduction of 121 EF2000 Eurofighter Typhoons, replacing the leased F-16 Fighting Falcons. Further updates are foreseen in the Tornado IDS/IDT and AMX fleets. A transport capability is guaranteed by a fleet of 22 C-130Js and Aeritalia G.222s of which 12 are being replaced with the newly developed G.222 variant called the C-27J Spartan.

Gendarmerie

An autonomous corps of the military, the Carabinieri are the gendarmerie and military police of Italy, policing the military and civilian population alongside Italy's other police forces. While the different branches of the Carabinieri report to separate ministries for each of their individual functions, the corps reports to the Ministry of Internal Affairs when maintaining public order and security. At the Sea Islands Conference of the G8 in 2004, the Carabinieri were given the mandate to establish a Center of Excellence for Stability Police Units (CoESPU) to spearhead the development of training and doctrinal standards for civilian police units attached to international peacekeeping missions.

Italy is subdivided into 20 regions (regioni, singular regione). Five of these regions have a special autonomous status that enables them to enact legislation on some of their local matters; these are marked by an asterisk (*) in the table below. The country is further divided into 110 provinces (province) and 8,100 municipalities (comuni).

Demographics

Population

At the end of 2008, the Italian population surpassed 60 million. Italy has the fourth-largest population in the European Union and the 23rd-largest population worldwide. Italy's population density, at 199.2 persons per square kilometre, is the fifth highest in the European Union.

The highest density is in Northern Italy, as that one-third of the country contains almost half of the total population. After World War II, Italy enjoyed a prolonged economic boom which caused a major rural exodus to the cities, and at the same time transformed the nation from a massive emigraticn country to a net immigrant-receiving country. High fertility persisted until the 1970s, when it plunged below the replacement rates, so that as of 2008, one in five Italians was over 65 years old.

Despite this, thanks mainly to the massive immigration of the last two decades, in the 2000s, Italy experienced a growth in the crude birth rate (especially in the northern regions) for the first time in many years. The total fertility rate has also significantly grown in the past few years, thanks to rising births among both in foreign-born and Italian women, as it climbed from 1.32 children per woman in 2005 to 1.41 in 2008.

Immigration

At the start of 2010 there were 4,279,000 foreign nationals resident in Italy and registered with the authorities. This amounted to 7.1% of the country's population and represented a year-on-year increase of 388,000. These figures include more than half a million children born in Italy to foreign nationals—second generation immigrants are becoming an important element in the demographic picture—but exclude foreign nationals who have subsequently acquired Italian nationality; this applied to 53,696 people in 2008. They also exclude illegal immigrants, the so-called clandestini whose numbers are difficult to determine. In May 2008 The Boston Globe quoted an estimate of 670,000 for this group.

Since the expansion of the European Union, the most recent wave of migration has been from surrounding European nations, particularly Eastern Europe, and increasingly Asia, replacing North Africa as the major immigration area. Some 950,000 Romanians, around 10 percent of them being Gypsies, are officially registered as living in Italy, replacing Albanians and Moroccans as the largest ethnic minority group. The number unregistered Romanians is difficult to estimate, but the Balkan Investigative Reporting Network suggested that in 2007 that there might half been half a million or more.

Crime

The Mafia originates from Italy and its influence is widespread in Italian society, directly affecting 22% of Italians and 14.6% of Italy's GDP, while even Prime Minister Silvio Berlusconi has long been accused of links with organised crime. The fight against the Mafia has cost the lives of many, including the high profile assassinations of judges Giovanni Falcone and Paolo Borsellino.

There are four separate Mafias controlling territory and business activities in four Southern Italian regions: Cosa Nostra in Sicily, Camorra in Campania, 'Ndrangheta in Calabria and Sacra Corona Unita in Puglia, exerting influence over 13 million Italians. Their business involvement reaches European and global scale. Businesses, enterpreneurs, shopkeepers and craftsmen in these regions are expected to pay a "pizzo", or protection money, to crime syndicates controlling their area. There rarely is possibility of escaping payment, and persons not complying find their business premises and their lives at risk. Those not able to meet demands might find their business partly or completely taken over by organised crime.

At 0.013 per 1,000 people, Italy has the 47th highest murder rate in the world, in a sample of 63 countries.. It also has the 43th highest number of rapes/1.000 people in the world, in a sample of 65 countries.

Economy

According to the International Monetary Fund, in 2008 Italy was the seventh-largest economy in the world and the fourth-largest in Europe. Italy is member of the Group of Eight (G8) industrialized nations, the European Union and the OECD.

In the post-war period, Italy was transformed from a weak, agricultural based economy which had been severely affected by the consequences of World War II, into one of the world's most industrialized nations and a leading country in world trade and exports, even so that in 1987, the Italian economy surpassed the British economy, by GDP (nominal), an event known as 'il sorpasso'.

According to the World Bank, Italy has high levels of freedom for investments, business and trade. Italy is a developed country, and, according to The Economist, has the world's 8th highest quality of life. The country enjoys a very high standard of living, and is the world's 18th most developed country, surpassing Germany, the UK and Greece. According to the last Eurostat data, Italian per capita GDP at purchasing power parity remains approximately equal to the EU average.

In addition, Italy has the world's 4th (3rd excluding the IMF) largest gold reserves, that of 2,451.8 tonnes, coming after the USA and Germany, and surpassing France and China. The country is also well-known for its influential and innovative business economic sector, an industrious and competitive agricultural sector, and for its creative and high-quality automobile, industrial, appliance and fashion design. Despite this, the country's economy suffers from many problems. After a strong GDP growth of +8% from 1964 onwards, the last decade's average annual growth rate lagged with 1.23% in comparison to an average EU annual growth rate of 2.28%. In addition, Italian living standards have a considerable north-south divide. The average GDP per capita in Northern Italy can far exceed the EU average (an example of this could be the Province of Bolzano-Bozen, with a 2006 average GDP per capita of €32,900 (US$ 43,861), which is 135.5% of EU average), whilst some regions and provinces in Southern Italy can be considerably below the EU average (such as Campania, which has an average GDP per capita of € 16,294, or US$ 21,722). Italy has often been referred the sick man of Europe, characterised by economic stagnation, political instability and problems in pursuing reform programs.

Italy suffers from structural weaknesses due to its geographical conformation and the lack of raw materials and energy resources. The territory is mostly mountainous, so much of the terrain is not suitable for intensive cultivation and communication is made more difficult.

The Italian economy is weakened by the lack of infrastructure development, market reforms and research investment, and also high public deficit. In the Index of Economic Freedom 2008, the country ranked 64th in the world and 29th in Europe, the lowest rating in the Eurozone. Italy still receives development assistance from the European Union every year. Between 2000 and 2006, Italy received €27.4 billion from the EU. The country has an inefficient state bureaucracy, low property rights protection and high levels of corruption, heavy taxation and public spending that accounts for about half of the national GDP. In addition, the most recent data show that Italy's spending in R&D in 2006 was equal to 1.14% of GDP, below the EU average of 1.84% and the Lisbon Strategy target of devoting 3% of GDP to research and development activities.

Italy has a smaller number of world-class multinational corporations than other economies of comparable size, but there are a large number of small and medium-sized enterprises, and in the Northern "industrial triangle" (Milan-Turin-Genoa) or the Tuscan industrial triangle (Florence-Prato-Pistoia), where there is an area of intense industrial and machinery production, notably in their several industrial districts, which were for long the backbone of the Italian industry. This has produced a manufacturing sector often focused on the export of niche market and luxury products, capable of facing the competition from China and other emerging Asian economies based on lower labour costs.

Italy's major exports and companies by sector are motor vehicles (Fiat Group, Aprilia, Ducati, Piaggio); chemicals and petrochemicals (Eni); energy and electrical engineering (Enel, Edison); home appliances (Candy, Indesit), aerospace and defense technologies (Alenia, Agusta, Finmeccanica), firearms (Beretta), fashion (Armani, Valentino, Versace, Dolce & Gabbana, Roberto Cavalli, Benetton, Prada, Luxottica); food processing (Ferrero, Barilla Group, Martini & Rossi, Campari, Parmalat); sport and luxury vehicles (Ferrari, Maserati, Lamborghini, Pagani); yachts (Ferretti, Azimut).

The automobile industry in Italy is also quite large employer in the country, with a labour force of over 196,000 (2004) working in this particular industry. The automotive industry makes a significant contribution of 8.5% to Italian GDP. Italy is the 5th largest automobile producer in Europe (2006).

Tourism

Tourism is one of the fastest growing and profitable sectors of the national economy: with 43.7 million international tourist arrivals and total receipts estimated at $42.7 billion, Italy is the fourth highest tourism earner and the fifth most visited country in the world. Despite a slump in the late-1980s and during the Gulf War, Italy has, since the mid-1990s, rebuilt a strong tourism industry. Italy's most popular tourist attractions are the Colosseum (4 million tourists per year, and the world's 39th most visited sight) and the ruins at Pompeii (48th in the world, with 2.5 million visitors).

In 2008, Italy's most internationally visited cities were in order: Rome (11th in the world with 6,123,000 tourists), Milan (52nd with 1,914,000 visitors), Venice (57th with 1,798,000), Florence (59th in the world with 1,729,000 international arrivals), Naples (166th with 381,000 tourists), Palermo (183rd with 316,000 visitors), Verona (188th with 289,000), Rimini

(189th with 284,000 international visitors), Bologna (191st in the world, with 279,000 arrivals), Genoa (200th, with 243,000 tourists), Turin (203rd with 240,000 visitors), and finally Siena (229th, with 163,000).

Infrastructure

In 2004 the transport sector in Italy generated a turnover of about 119.4 billion euros, employing 935,700 persons in 153,700 enterprises. Regarding the national road network, in 2002 there were 668,721 km (415,612 mi) of serviceable roads in Italy, including 6,487 km (4,031 mi) of motorways, state-owned but privately operated by Atlantia. In 2005, about 34,667,000 passenger cars (590 cars per 1,000 people) and 4,015,000 goods vehicles circulated on the national road network.

The national railway network, state-owned and operated by Ferrovie dello Stato, in 2003 totalled 16,287 km (10,122 mi) of which 69% is electrified, and on which 4,937 locomotives and railcars circulated. The national inland waterways network comprised 1,477 km (918 mi) of navigable rivers and channels in 2002. In 2004 there were approximately 30 main airports (including the two hubs of Malpensa International in Milan and Leonardo Da Vinci International in Rome) and 43 major seaports (including the seaport of Genoa, the country's largest and second largest in the Mediterranean Sea). In 2005 Italy maintained a civilian air fleet of about 389,000 units and a merchant fleet of 581 ships. Italy's economy has a diversified energy sector highly dependent on imports: in 2006 the country imported more than 86% of its total energy consumption (99.7% of the solid fuels, 92.5% of oil, 91.2% of natural gas and 15% of electricity) Italy, however, ranked as the world's sixth largest producer of wind power behind India and ahead of France and the United Kingdom, with an installed nameplate capacity of 3,736 GW in 2008. There are no active nuclear powerplants within the country. The majority of Italy's electricity is produced by gas, oil, coal, and hydroelectric.

Exports

The country is the world's 8th largest exporter. In 2007, Italy's total merchandise exports were at US$ 491.507 billion,

US$ 118.261 billion for commercial services and they were at US$ 504.404 billion by selected region, according to the World Trade Organization.

Italy's major exports are precision machinery, motor vehicles (utilitaries, luxury vehicles, motorcycles, scooters), chemicals and electric goods, but the country's more famous exports are in the fields of food and clothing.

Italy is an important agricultural exporter, being the largest worldwide exporter and producer of kiwis, grape (8,519,418 tonnes) and artichoke (469,980 tonnes). Italy was responsible for producing approximately one-fifth of world wine production in 2005. Italy's closest trade ties are with the other countries of the European Union, with whom it conducts about 59% of its total trade. Its largest EU trade partners, in order of market share, are Germany (12.9%), France (11.4%), and Spain (7.4%).

Public Services

Healthcare

Italy has a nationally run healthcare system which was founded in 1978. Healthcare spending in Italy has accounted for more the 9.0% of the country's GDP, slightly above the OECD countries' average of 8.9%, however, this has resulted in Italy having the world's 2nd best healthcare system, 19th highest life expectancy, and the world's 3rd best healthcare performance. Italy's life expectancy at birth was in 2004 80.9, two years above the OECD average. The infant mortality rate in Italy was 4.7 % (OECD: 5.4 %) in 2005.

Nevertheless, in Italy, just like in virtually all OECD countries, there has been an increase in the proportion of overweight and obese people. This figure rose from 7.0% in 1994 to 9.9% in the year 2005. There are now also in hospital structures in Italy for this group of people

In addition, the proportion of daily smokers fell in the same period from 1990 to 2005 from 27.8% to 22.3% (OECD: 24.3%). Since 10 January 2005, there is also a general smoke ban in all public buildings. Anyone who violates this law must pay a fine of €27.5 to €275.

Education

Italy's public education is free and compulsory from 6–14 years of age, and has a five-year primary stage and an eight-year secondary stage, divided into first-grade secondary school (middle school) and second-grade secondary school (or high school). Italy has a high public education standard, surpassing that of other comparable developed countries, such as the UK and Germany. The country has both public and private education systems.

According to National Science Indicators (1981–2002), a database produced by Research Services Group containing listings of output and citation statistics for more than 90 countries, Italy has an above-average output of scientific papers (in terms of number of papers written with at least one author being from Italy) in space science (9.75% of papers in the world being from Italy), mathematics (5.51% of papers in the world), computer science, neurosciences, and physics; the lowest, but still slightly above world-average, output in terms of number of papers produced is recorded in the social sciences, psychology and psychiatry, and economics and business.

Turin's Castello del Valentino, which is the seat of the Polytechnic University of Turin.

Italy hosts a broad variety of universities, colleges and academies. Milan's Bocconi University, has been ranked among the top 20 best business schools in the world by The Wall Street Journal international rankings, especially thanks to its M.B.A. program, which in 2007 placed it no. 17 in the world in terms of graduate recruitment preference by major multinational companies. Also, Forbes has ranked Bocconi no.1 worldwide in the specific category Value for Money. In May 2008, Bocconi overtook several traditionally top global business schools in the Financial Times Executive education ranking, reaching no. 5 in Europe and no. 15 in the world.

Other top universities and polytechnics include the Polytechnic University of Turin, the Politecnico di Milano (which in 2009 was ranked as the 57th technical university in the world by Top Universities, in a research conducted on behalf of Times Higher Education. This was a 6-positions growth from the 63rd position in 2008. In 2009 an Italian research ranked it as the best in Italy over indicators such as scientific production, attraction of foreign students, and others), the La Sapienza (which in 2005 was Europe's 33rd best university, and ranks amongst Europe's 50 and the world's 150 best colleges) and the University of Milan (whose research and teaching activities have developed over the years and have received important international recognitions. The University is the only Italian member of the League of European Research Universities (LERU), a prestigious group of twenty research-intensive European Universities. It also been awarded ranking positions as such:-1st in Italy and 7th in Europe (The Leiden Ranking-Universiteit Leiden).

Italy and the Western world's oldest college is the University of Bologna. In 2009, the University of Bologna is, according to The Times, the only Italian college in the top 200 World Universities. The University of Padua, also remains one of Europe's oldest.

Communications

Italy has modern telephone and data services. The country has 17.7 million internet hosts, 4th-most in the world, and 32 million internet users, 10th highest in the world. There are 88.58 million mobile cellular telephones in Italy, far exceeding the actual population and ranking 11th in the world, and 20 million landline telephones. Italy has high-capacity cables for domestic usage of phones, and numerous international connections.

Media and Censorship

The first form of televised media in Italy was introduced in 1939, when the first experimental broadcasting began. However, this lasted for a very short time: when fascist Italy entered World War II in 1940 all the transmission were interrupted, and were resumed in earnest only nine years after the end of the conflict, in 1954. There are two main national television networks responsible for most viewing: state-owned RAI, funded by a yearly mandatory licence fee and Mediaset, commercial network founded by current Italian Prime Minister

Silvio Berlusconi. While many other networks are also present, both nationally and locally, these two alone account for 80% of the TV ratings.

As with all the other media of Italy, the Italian television industry is widely considered both inside and outside the country to be overtly politicized. The public broadcaster RAI is, unlike the BBC which is controlled by an independent trust, under direct control of the government; the most important commercial stations in the country are, in turn, owned by the current prime minister. According to a December 2008 poll, only 24% of Italians trust television news programmes, comparing unfavourably to the British rate of 38%, making Italy one of only three examined countries where online sources are considered more reliable than television ones for information. Along with Turkey, Italy has one of the lowest levels of press freedom in Europe, even falling behind some ex-communist countries, such as Poland and the Czech Republic.

Society

Italy has been nominated 2009's sixth most internationally valued country, (falling short of France, Germany, The United Kingdom, Canada and Japan, and surpassing the United States, Switzerland, Australia, Sweden and Spain), coming ninth in export branding 2008, first in tourism branding, second in cultural branding, third in people branding and ninth in immigration branding.

Social Class

Italian society is often divided into different ranks of social class. Italian society is divided between the Bourgeosie, the white collar middle class, the petit urban bourgeosie, the petit rural bourgeosie, the urban working class and the rural working class.

Women

Women have equal rights as men, and have mainly the same job, business and education opportunities. Some traditionalist people in Italian society (especially in the South) still tend to treat women as slightly inferior, but women's rights in Italy are comparable to most western countries.

LGBT Rights

Italy legally accepts homosexuals and transgenders, however they may face legal challenges not experienced by non-LGBT residents. Both male and female same-sex sexual activity are legal in Italy, but same-sex couples and households headed by same-sex couples are not eligible for the same legal protections available to opposite-sex couples.

Italian opinions have changed in the past and people now tend to be more supportive and liberal of lesbian, gay, bisexual, transgender (LGBT) rights, but tend to be more repressive than other European nations. Tolerance is seen in a peculiar way that is due to the religious influence of the Roman Catholic Church, which has been ingrained in Italian society for 1,700 years. Conservative Italian politicians such as Silvio Berlusconi have often been opposed to increasing gay rights. A Eurobarometer survey published on December 2006 showed that 31% of Italians surveyed support same-sex marriage and 24% recognise same-sex couple's right to adopt (EU-wide average 44% and 33%). A recent 2007 poll asking whether they supported the civil partnership law for gays. Support for the measure was at 45% support, with 47% oppose. 8% said they were unsure. Homosexuals are also allowed to serve fully in military service.

Daily Life and Leisure

Italians' social customs and daily lives have profoundly changed since World War II, transforming the nation from a highly traditional, agricultural-based society, into a progressive and modernized one.

Most Italians favour activities such as going to the cinema, reading newspapers, watching television and listening to the radio; reading books and playing sport has proved less popular. According to some surveys, Italians are generally highly satisfied with social relations and family, healthcare, daily life and friendship relations; however, Italians find economic status and job opportunites generally less satisfying, especially in parts of Southern Italy that suffer from relatively high unemployment. Meeting up and socializing with friends in the country's abundant piazzas, going to bars, discos, pizzerias and restaurants and finding other forms of entertainment remain popular with Italians, especially the younger generations. Automobiles still hold a strong part of Italian daily life, however this results in many cities being congested.

Customs and Etiquette

Italy has a code of etiquette comparable to most etiquette values in Europe. Being open and confident and presenting oneself well by dressing elegantly when meeting new people is considered essential for what is called la bella figura, making a good impression. Dining is considered important, with table manners being highly valued. Families tend to eat together around the table, and on Sundays or special occasions people may invite friends or relatives home or out to a restaurant to have a meal together.

Environment

After its quick industrial growth, Italy took a long time to confront its environmental problems. After several improvements, it now ranks 84th in the world for ecological sustainability. Air pollution is a severe problem, especially in the industrialised north, reaching the tenth worldwide highest level of industrial carbon dioxide emissions in the 1990s (7.03 tons per capita). Carbon monoxide emissions in the country increased by 12% in the period between 1985 and 1989.

Extensive traffic and congestion in the bigger metropolitan areas of Italy, such as Milan, Rome and Naples has meant that pollution levels in Italian urban areas tends to be high, causing environmental and health issues. The problem has been recently addressed by 80 towns and cities of Northern Italy during an anti-smog day in which all non-essential traffic was halted. Smog levels have decreased dramatically since the 1970s and 80s, and the presence of smog is becoming an increasingly rarer phenomenon and levels of sulphur dioxide are decreasing. The air in less built-up areas, such as the Italian Alps and parts of Southern Italy, tends to be far less polluted than in industrialised areas.

Many watercourses and coastal stretches have also been contaminated by industrial and agricultural activity, while due to rising water levels Venice has been regularly flooded throughout recent years. Waste from industrial activity is not always disposed of by legal means and has led to permanent health effects on inhabitants of affected areas, as in the case of the Seveso disaster. The country has also operated several nuclear reactors between 1963–1990 but, after the Chernobyl disaster and a referendum on the issue the nuclear program was terminated, a decision that was overturned by the government in 2008. A deal was signed with France in 2009 for the construction of up to four new nuclear plants. See Nuclear power in Italy.

Deforestation, illegal building developments and poor land management policies have led to significant erosion all over Italy's mountainous regions, leading to occasional disasters and major loss of life like the Vajont Dam flood, the Messina mudslides, and the Sarno mudslide that killed 137 people in 1998.

Culture

Italy did not exist as a state until the country's unification in 1861. Due to this comparatively late unification, and the historical autonomy of the regions that comprise the Italian Peninsula, many traditions and customs that are now recognized as distinctly Italian can be identified by their regions of origin. Despite the political and social distinction of these regions, Italy's contributions to the cultural and historical heritage of Europe and the world remain immense. Italy is home to the greatest number of UNESCO World Heritage Sites to date, and has rich collections of world art, culture and literature from many different periods. Italy has had a broad cultural influence worldwide, also due to the fact that numerous Italians emigrated to other countries during the Italian diaspora. Italy has, overall, an estimated 100,000 monuments of any sort (museums, palaces, buildings, statues, churches, art galleries, villas, fountains, historic houses and archaeological remains).

Arts

Italy boasts a long period of different architectural styles, from Classical Roman and Greek, Gothic, Renaissance, Baroque, Neo-Classical, Art Nouveau to Modern. The nation contains several architectural monuments, such as the Pantheon, the Colosseum, the Leaning Tower of Pisa, the Piazza del Campo, Milan Cathedral, Florence Cathedral, the Palladian Villas of the Veneto, the Basilica di Santa Maria Maggiore, Villa Olmo and the Pirelli Tower. Italy has also been home to numerous famous architects, some who even changed the course of architectural history, such as Andrea Palladio (who founded Palladianism), Filippo Brunelleschi, Bernini and Renzo Piano.

Classical to Gothic

Italian architecture began with Ancient Greece, Ancient Rome and Etruscans, when both civilizations built temples, basilicae, columns, fora, palaces, aqueducts, walls and public baths. Roman architecture had great influence on that of Italy and the Western world. Because the Roman Empire extended over so great an area and included so many urbanised areas, Roman engineers developed methods for civic development on a grand scale, including the use of concrete. Massive buildings like the Pantheon and the Colosseum could never have been constructed with pre-existing techniques. Though concrete had been invented a thousand years earlier in the Near East, the Romans extended its use from fortifications to their most impressive buildings and monuments, capitalising on the material's strength and low cost. In Roman architecture, a wall's concrete core was covered with a plaster, brick, stone, or marble veneer, and decorative polychrome and gold-gilded sculpture was often added to produce a dazzling effect of power and wealth. Gothic architecture appeared in Italy in the 12th century, but did not mature into a regionally distinct style until the 13th century, partly due to geographic factors. Due to its comparatively late maturity, the influence of Byzantine and classical art, and the fact that brick —not stone— was the most common building material and marble the most common decorative material, Italian Gothic architecture maintained peculiar characteristics which differentiated its evolution from that in France, where it had originated, and in other European countries.

In particular, the architecturally daring solutions and technical innovations of the French Gothic cathedrals rarely appeared. With the exception of the Cathedral of Milan, the product of a centuries-long collaboration between Italian, French, and German minds, few Italian churches show the emphasis on vertical development, clustered shafts, ornate tracery and complex ribbed vaulting that characterise Gothic in other parts of Europe. No examples of Italian Gothic architecture include Basilica of Santa Croce, Orvieto Cathedral, and Siena Cathedral, where the distinctively ornate Italian realization of facade design is evident.

Renaissance to Modern

Italy of the 15th century, and the city of Florence in particular, was home to the Renaissance. It is in Florence that the new architectural style had its beginning, not slowly evolving in the way that Gothic grew out of Romanesque, but consciously brought to being by particular architects who sought to revive the order of a past "Golden Age". The scholarly approach to the architecture of the ancient coincided with the general revival of learning. A number of factors were influential in bringing this about.

Italian architects had always preferred forms that were clearly defined and structural members that expressed their purpose. Many Tuscan Romanesque buildings demonstrate these characteristics, as seen in the Florence Bapistry and Pisa Cathedral.

The presence, particularly in Rome, of ancient architectural remains showing the ordered Classical style provided an inspiration to artists at a time when philosophy was also turning towards the Classical.

Italy then became a main European centre for the baroque, with diverse baroque architectural styles emerging, especially in Sicily. In the 18th and 19th centuries neo-classical style buildings began to appear in Rome, Milan, Turin and all around Italy. The 19th century also saw the construction of several considerable works of Italian architecture, including the Galleria Vittorio Emanuele II in Milan, one of the world's oldest shopping galleries, which influenced others such as the Galleria Umberto I in Naples, the Burlington Arcade in London and the Passazh in Saint Petersburg.

In the 20th century, Italy too saw the construction of several significant edifices, starting in the Art Nouveau architectural style, which in Italy was called, Liberty architecture. Rationalist-Fascist architecture developed in Italy during the Fascist era, lasting until the 1940s. During that period Italy built the world's first motorway between Milan and Varese in 1921, and considerable architectural works of the era include Fiat's Lingotto, at the time the world's biggest automobile factory. In the 1950s and 60s several skyscrapers were built across the country, the Pirelli Tower and the Torre Velasca being the most notable. The 21st century most notable Italian buildings are at the Fiera Milano exposition centre in Rho, just outside Milan, (one of Europe's biggest and most important exhibition centres) and the new plans for the Expo 2015 to be held in Milan too, where three new skyscapers called "lo storto", "il curvo" and "il diritto" will be constructed by foreign architects such as Zaha Hadid, Arata Isozaki and Daniel Libeskind. This will also be a project or urban re-development, called "City-Life", where new pedestrian areas, parks, green spaces, lakes and waterways will be constructed, in the North-Western part of Milan.

Palazzi and Villas

Italy boasts a wide variety of palaces, in various cities, mainly Rome, Florence, Venice, Milan, Turin, Bologna and Naples, built in a wide variety of different styles, from Roman, Byzantine, Romanesque, Medieval and Gothic, to Renaissance, Baroque, Rococo, Neo-Classical and Fascism. In Italian, the word "Palazzo" is more broadly used in Italy than its English equivalent "palace". In Italy, a palazzo is a grand building of some architectural ambition that is the headquarters of a family of some renown or of an institution, or even what the British would call a "block of flats" or a tenement. In Venice, most palaces are referred to as "Ca'", which is short for "Casa", meaning "house" in Italian, for example Ca' Pesaro or Ca' Rezzonico.

Gardens and Villas

Italy has several villas with notable formal gardens, most of which designed as Italian gardens, such as those of Villa d'Este. Their principles are of perfect geometry and symmetry and of imposing order over nature. Italian gardens were influenced by Roman gardening and Italian Renaissance gardening, and have been copied by other courts around Europe over the centuries.

The Italian Renaissance garden emerged in the late fifteenth century at villas in Rome and Florence. Inspired by classical ideals of order and beauty, they were intended to provide a pleasurable view of the garden and the landscape beyond it, for contemplation, and for enjoyment of the sights, sounds and smells of the garden itself.

In the late Renaissance, the gardens became larger, grander and more symmetrical, and were filled with fountains, statues, grottoes, water organs and other features designed to delight their owners and amuse and impress visitors. The style was imitated throughout Europe, influencing the gardens of the French Renaissance and the English garden.

Visual Art

Over the centuries, Italian art has gone through many stylistic changes. Italian painting is traditionally characterized by a warmth of colour and light, as exemplified in the works of Caravaggio and Titian, and a preoccupation with religious figures and motifs. Italian painting enjoyed pre-eminence in Europe for hundreds of years, from the Romanesque and Gothic periods, and through the Renaissance and Baroque periods, the latter two of which saw fruition in Italy. Notable artists who fall within these periods include Michelangelo, Leonardo da Vinci, Donatello, Botticelli, Fra Angelico, Tintoretto, Caravaggio, Bernini, Titian and Raphael.

Thereafter, Italy was to experience a continual subjection to foreign powers which caused a shift of focus to political matters, leading to its decline as the artistic authority in Europe. Not until 20th century Futurism, primarily through the works of Umberto Boccioni and Giacomo Balla, would Italy recapture any of its former prestige as a seminal place of artistic evolution. Futurism was succeeded by the metaphysical paintings of Giorgio de Chirico, who exerted a strong influence on the Surrealists and generations of artists to follow.

Literature

The basis of the modern Italian language was established by the Florentine poet Dante Alighieri, whose greatest work, the Divine Comedy, is considered amongst the foremost literary statements produced in Europe during the Middle Ages. There is no shortage of celebrated literary figures in Italy: Giovanni Boccaccio, Giacomo Leopardi, Alessandro Manzoni, Torquato Tasso, Ludovico Ariosto, and Petrarch, whose best-known vehicle of expression, the sonnet, was invented in Italy.

Prominent philosophers include Giordano Bruno, Marsilio Ficino, Niccolo Machiavelli, and Giambattista Vico. Modern literary figures and Nobel laureates are nationalist poet Giosue Carducci in 1906, realist writer Grazia Deledda in 1926, modern theatre author Luigi Pirandello in 1936, poets Salvatore Quasimodo in 1959 and Eugenio Montale in 1975, satirist and theatre author Dario Fo in 1997.

Theatre

Italian theatre can be traced back to the Roman tradition which was heavily influenced by the Greek; as with many other literary genres, Roman dramatists tended to adapt and translate from the Greek. For example, Seneca's Phaedra was based on that of Euripides, and many of the comedies of Plautus were direct translations of works by Menander. During the 16th century and on into the 18th century, Commedia dell'arte was a form of improvisational theatre, and it is still performed today. Travelling troupes of players would set up an outdoor stage and provide amusement in the form of juggling, acrobatics, and, more typically, humorous plays based on a repertoire of established characters with a rough storyline, called canovaccio.

Music

From folk music to classical, music has always played an important role in Italian culture. Instruments associated with classical music, including the piano and violin, were invented in Italy, and many of the prevailing classical music forms, such as the symphony, concerto, and sonata, can trace their roots back to innovations of sixteenth-and seventeenth-century Italian music.

Italy's most famous composers include the Renaissance composers Palestrina and Monteverdi, the Baroque composers Alessandro Scarlatti, Corelli and Vivaldi, the Classical composers Paganini and Rossini, and the Romantic composers Verdi and Puccini. Modern Italian composers such as Berio and Nono proved significant in the development of experimental and electronic music. While the classical music tradition still holds strong in Italy, as evidenced by the fame of its innumerable opera houses, such as La Scala of Milan and San Carlo of Naples, and performers such as the pianist Maurizio Pollini and the late tenor Luciano Pavarotti, Italians have been no less appreciative of their thriving contemporary music scene.

Italy is widely known for being the birthplace of opera. Italian opera was believed to have been founded in the early 1600s, in Italian cities such as Mantua and Venice. Later, works and pieces composed by native Italian composers of the 19th and early 20th centuries, such as Rossini, Bellini, Donizetti, Verdi and Puccini, are amongst the most famous operas ever written and today are performed in opera houses across the world. La Scala operahouse in Milan is also renowned as one of the best in the world. Famous Italian opera singers include Enrico Caruso, Alessandro Bonci, the late Luciano Pavarotti, and Andrea Bocelli, to name a few.

Introduced in the early 1920s, jazz took a particularly strong foothold in Italy, and remained popular despite the xenophobic cultural policies of the Fascist regime. Today, the most notable centres of jazz music in Italy include Milan, Rome, and Sicily. Later, Italy was at the forefront of the progressive rock movement of the 1970s, with bands like PFM and Goblin. Today, Italian pop music is represented annually with the Sanremo Music Festival, which served as inspiration for the Eurovision song contest, and the Festival of Two Worlds in Spoleto. Singers such as pop diva Mina, classical crossover artist Andrea Bocelli, Grammy winner Laura Pausini, and European chart-topper Eros Ramazzotti have attained international acclaim.

Cinema

The history of Italian cinema began a few months after the

Lumiere brothers began motion picture exhibitions. The first Italian film was a few seconds long, showing Pope Leo XIII giving a blessing to the camera. The Italian film industry was born between 1903 and 1908 with three companies: the Society Italiana Cines, the Ambrosio Film and the Itala Film. Other companies soon followed in Milan and in Naples. In a short time these first companies reached a fair producing quality, and films were soon sold outside Italy. Cinema was later used by Benito Mussolini, who founded Rome's renowned Ciencifica studio for the production of Fascist propaganda until the World War II.

After the war, Italian film was widely recognised and exported until an artistic decline around the 1980s. Notable Italian film directors from this period include Vittorio De Sica, Federico Fellini, Sergio Leone, Pier Paolo Pasolini, Michelangelo Antonioni and Dario Argento. Movies include world cinema treasures such as La dolce vita, Il buono, il brutto, il cattivo and Ladri di biciclette. In recent years, the Italian scene has received only occasional international attention, with movies like La vita e bella directed by Roberto Benigni and Il postino with Massimo Troisi.

Science and Technology

Through the centuries, Italy has given birth to some notable scientific minds. Amongst them, and perhaps the most famous polymath in history, Leonardo da Vinci made several contributions to a variety of fields including art, biology, and technology. Galileo Galilei was a physicist, mathematician, and astronomer who played a major role in the Scientific Revolution. His achievements include improvements to the telescope and consequent astronomical observations, and support for Copernicanism. The physicist Enrico Fermi, a Nobel prize laureate, was the leader of the team that built the first nuclear reactor and is also noted for his many other contributions to physics, including the co-development of the quantum theory.

A brief overview of some other notable figures includes the astronomer Giovanni Domenico Cassini, who made many important discoveries about the Solar System; the physicist Alessandro Volta, inventor of the electric battery; the mathematicians Lagrange, Fibonacci, and Gerolamo Cardano, whose Ars Magna is generally recognized as the first modern treatment on mathematics, made fundamental advances to the field; Marcello Malpighi, a doctor and founder of microscopic anatomy; the biologist Lazzaro Spallanzani, who conducted important research in bodily functions, animal reproduction, and cellular theory; the physician, pathologist, scientist, and Nobel laureate Camillo Golgi, whose many achievements include the discovery of the Golgi complex, and his role in paving the way to the acceptance of the Neuron doctrine; and Guglielmo Marconi, who received the Nobel Prize in Physics for the invention of radio.

Sport

The sport with the highest spectator attendance and number of registered players is football, the Serie A being one of the most famous competitions in the world. Italy's national football team is the second-most-successful team in the world, with four World Cup victories, the first one of which was in 1934.

Other popular sports include basketball, volleyball, waterpolo, Rugby Union, Rugby League, cycling, fencing, ice hockey (mainly in Milan, Trentino-Alto Adige and Veneto), roller hockey, motor racing and swimming. Winter sports are most popular in the northern regions, with Italians competing in international games and Olympic venues. Turin hosted the 2006 Winter Olympic Games.

Numerous folk sports are played around the country, sometimes incorporated into Italian festivities. Examples are Palio horse racing (the most renowned being the Palio di Siena) and traditional regattas like the gondola regatta in Venice and the annual regatta of the Maritime Republics, and reenactments of ancient sports like Calcio fiorentino.

Others folk sports played regularly or on a league base include Tamburello, Bocce and Boccia su strada, a game similar to Klootschieten.

Fashion

Italian fashion has a long tradition, and is regarded as one of the most important in the world, along with French fashion,

American fashion, British fashion and Japanese fashion. Milan, Florence and Rome are Italy's main fashion capitals, however Naples, Turin, Venice, Bologna, Genoa and Vicenza are other major centres. According to the 2009 Global Language Monitor, Milan was nominated the true fashion capital of the world, even surpassing other international cities, such as New York, Paris, London and Tokyo, and Rome came 4th. Major Italian fashion labels, such as Gucci, Prada, Versace, Valentino, Armani, Dolce & Gabbana, Missoni, Fendi, Moschino, Max Mara and Ferragamo, to name a few, are regarded as amongst the finest fashion houses in the world. Also, the fashion magazine Vogue Italia, is considered the most important and prestigious fashion magazine in the world.

Design

Italy is prominent in the field of design, notably interior design, architectural design, industrial design and urban design. Italy has produced some well-known furniture designers, such as Gio Ponti and Ettore Sottsass, and Italian phrases such as "Bel Disegno" and "Linea Italiana" have entered the vocabulary of furniture design. Examples of classic pieces of Italian white goods and pieces of furniture include Zanussi's washing machines and fridges, the "New Tone" sofas by Atrium, and the post-modern bookcase by Ettore Sottsass, inspired by Bob Dylan's song Memphis Blues. Today, Milan and Turin are the nation's leaders in architectural design and industrial design. The city of Milan hosts the Fiera Milano, Europe's biggest design fair. Milan also hosts major design and architecture-related events and venues, such as the "Fuori Salone" and the "Salone del Mobile", and has been home to the designers Bruno Munari, Lucio Fontana, Enrico Castellani and Piero Manzoni

Cuisine

Modern Italian cuisine has evolved through centuries of social and political changes, with its roots reaching back to the 4th century BC. Significant change occurred with the discovery of the New World, when vegetables such as potatoes, tomatoes, bell peppers, and maize became available. However, these central ingredients of modern Italian cuisine were not introduced in scale before the 18th century.

Ingredients and dishes vary by region. However, many dishes that were once regional have proliferated in different variations across the country. Cheese and wine are major parts of the cuisine, playing different roles both regionally and nationally with their many variations and Denominazione di origine controllata (regulated appellation) laws. Coffee, and more specifically espresso, has become highly important to the cultural cuisine of Italy. Some famous dishes and items include pasta, pizza, lasagna, focaccia, and gelato.

3

Spain and Portugal: The Iberian Peninsula

The Iberian Peninsula is located in the extreme southwest of Europe and includes modern-day states Portugal, Spain, Andorra, the British Overseas Territory of Gibraltar and a very small area of France. It is the westernmost of the three major southern European peninsulas—the Iberian, Italian, and Balkan peninsulas. It is bordered on the southeast and east by the Mediterranean Sea, and on the north, west and southwest by the Atlantic Ocean. The Pyrenees form the northeast edge of the peninsula, separating it from the rest of Europe. In the south, it approaches the northern coast of Africa. It is the second-largest peninsula in Europe, with an area of 582,860 square kilometres (225,040 sq mi).

Name

Greek Name

The English word *Iberia* was adapted from the use of the Ancient Greek word Eacnba (Ibcria) by the Greek geographers under the Roman Empire to refer to what is known today in English as the Iberian Peninsula. The name was not then used to mean a single political country or a population speaking a single language. Strabo's Iberia was delineated from Keltikc by the Pyrenees and included the entire land mass south (he mistakenly said west) of there. The Ancient Greeks discovered Iberia by voyaging westward. Hecataeus of Miletus was the first known to use the term around 500 BC. Herodotus of Halicarnassus says of the Phocaeans that "it was they who made the Greeks acquainted with... Iberia." According to Strabo prior historians used Iberia to mean the country "this side of the >acnio (Ibcros)" as far north as the Rhone river in France but currently they set the Pyrenees as the limit. Polybius respects that limit but identifies Iberia as the Mediterranean side as far south as Gibraltar, with the Atlantic side having no name. Elsewhere he says that Saguntum is "on the seaward foot of the range of hills connecting Iberia and Celtiberia."

Strabo refers to the Carretanians as people "of the Iberian stock" living in the Pyrenees, who are to be distinguished from either Celts or Celtiberians.

The word "Iberia" is not currently used, since it is often associated with some political points of view, implying the unification of Portugal and Spain.

Roman Names

When the Romans encountered the Greek geographers they used Iberia poetically and spoke of the Iberi, the population of Iberia. First mention was in 200 BC by the poet Quintus Ennius. The Romans had already had independent experience with the peoples on the peninsula during the long conflict with Carthage. The Roman geographers and other prose writers from the time of the late Roman Republic called the entire peninsula *Hispania*.

As they became politically interested in the former territories of Carthage the Romans came to use Hispania Citerior and Hispania Ulterior for "near" and "far Spain". Even at that time large sections of it were Lusitania (Portugal south of Douro river and Extremadura in western Spain), Gallaecia (Northern Portugal and Galicia in Spain), Celtiberia (central Spain), Baetica (Andalusia), Cantabria (northwest Spain) and the Vascones (Basques). Strabo says that the Romans use Hispania and Iberia synonymously, and distance them as near and far. He was living in a time when the peninsula was divided into Roman provinces, some belonging "to the people and the Senate" and some to "the Roman emperor." Baetia was distinguished by being the only one belonging "to the people." Whatever language may have been spoken on the peninsula soon gave way to Latin, except for Basque, protected by the Pyrenees.

Etymology

"Iberia" has always been associated with the Ebro river, Ibcros in ancient Greek and Ibcrus or Hibcrus in Latin. The association was so well known it was hardly necessary to state; for example, Ibcria was the country "this side of the Ibcrus" in Strabo. Pliny goes so far as to assert that the Greeks had called "the whole of Spain" Hiberia because of the river Hiberus. The river appears in the Ebro Treaty of 226 BC between Rome and Carthage, setting the limit of Carthaginian interest at the Ebro. The fullest description of the treaty, stated in Appian, uses Ibcrus. With reference to this border, Polybius states that the "native name" is Ibcr, apparently the original word, stripped of its Greek or Latin-os or-us termination.

The early range of these natives, stated by the geographers and historians to be from southern Spain to southern France along the Mediterranean coast, is marked by instances of a readable script expressing a yet unknown language, dubbed "Iberian." Whether this was the native name or was given to them by the Greeks for their residence on the Ebro remains unknown. Credence in Polybius imposes certain limitations on etymologizing: if the language remains unknown, the meanings of the words, including Iber, must remain unknown also.

Geography

Geography of Spain

Spain is located in southwestern Europe and comprises about 84 percent of the Iberian Peninsula. Its total area is 504,782 km^2 (194,897 sq mi) of which 499,542 km^2 (192,874 sq mi) is land and 5,240 km^2 (2,023 sq mi) is water., and the Atlantic coast is 710 km (441 mi) long. The Pyrenees mountain range, extends 435 km (270 mi) from the Mediterranean to the Bay of Biscay. In the extreme south of Spain lie the Straits of Gibraltar, which separate Spain and the rest of Europe from Morocco in north Africa; at its narrowest extent, Spain and Morocco are separated by only 13 km (8.1 mi).

Off the Iberian Peninsula there are several other Spanish areas: the Balearic Islands in the Mediterranean, the Canary Islands to the southwest, 108 km (67 mi) off northwest Africa, and five places of sovereignty (*plazas de soberania*) on and off the coast of Morocco: Ceuta, Melilla, Islas Chafarinas, Penon de Alhucemas, and Penon de Velez de la Gomera.

External Boundaries and Landform Regions

Most of Spain's boundaries are water: the Mediterranean Sea on the south and east from Gibraltar to the French border and the Atlantic Ocean on the northwest and southwest (in the south as the Golfo de Cadiz and in the north as the Bay of Biscay). Spain also shares land boundaries with France and Andorra along the Pyrenees in the northeast, with Portugal on the west, and with the small British Overseas Territory of Gibraltar near the southernmost tip. The affiliation of Gibraltar has continued to be a contentious issue between Spain and Britain and so is the sovereignty of the plazas de soberania, claimed by Morocco.

The majority of Spain's peninsular region consists of the Meseta Central, a highland plateau rimmed and dissected by mountain ranges. Other landforms include narrow coastal plains and some lowland river valleys, the most prominent of which is the Andalusian Plain in the southwest. The country can be divided into ten natural regions or subregions: the dominant Meseta Central, the Cantabrian Mountains (Cordillera Cantabrica) and the northwest region, the *Iberico* region, the Pyrenees, the *Penibetico* region in the southeast, the Andalusian Plain, the Ebro Basin, the coastal plains, the Balearic Islands, and the Canary Islands. These are commonly grouped into four types: the Meseta Central and associated mountains, other mountainous regions, lowland regions, and islands.

The Inner Plateau and Associated Mountains

The *Meseta Central* ("Inner Plateau") is a vast plateau in the heart of peninsular Spain, has elevations that range from 610 to 760 m. Rimmed by mountains, the Meseta Central slopes gently to the west and to the series of rivers that form some of the border with Portugal. The Sistema Central, described as the "dorsal spine" of the Meseta Central, divides the Meseta into northern and southern subregions, the former higher in elevation and smaller in area than the latter. The Sistema Central rims the capital city of Madrid with peaks that rise to 2,400 m north of the city and to lower elevations south of it. West of Madrid, the Sistema Central shows its highest peak, Pico Almanzor, of 2,592 m. The mountains of the Sistema Central, which continue westward into Portugal, display some glacial features; the highest of the peaks are snow-capped for most of the year. Despite their height, however, the mountain system does not create a major barrier between the northern and the southern portions of the Meseta Central because several passes permit road and railroad transportation to the northwest and the northeast.

The southern portion of the Meseta is further divided by twin mountain ranges, the Montes de Toledo running to the east and the Sierra de Guadalupe, to the west. Their peaks do not rise much higher than 1,500 m. With many easy passes, including those that connect the Meseta with the Andalusian Plain, the Montes de Toledo and the Sierra de Guadalupe do not present an obstacle to transportation and communication. The two mountain ranges are separated from the Sistema Central to the north by the longest river in the Iberian Peninsula: the Tagus River.

The mountain regions that rim the Meseta Central and are associated with it are the Sierra Morena, the Cordillera Cantabrica, and the Sistema Iberico. Forming the southern edge of the Meseta Central, the Sierra Morena merges in the east with the southern extension of the Sistema Iberico and reaches westward along the northern edge of the Rio Guadalquivir valley to join the mountains in southern Portugal. The massif of the Sierra Morena extends northward to the Rio Guadiana, which separates it from the Sistema Central. Despite their relatively low elevations, seldom surpassing 1,300 m, the mountains of the Sierra Morena are rugged.

The Cordillera Cantabrica, a limestone formation, runs parallel to, and close to, the northern coast near the Bay of Biscay. Its highest points are the Picos de Europa, surpassing 2,600 m. The Cordillera Cantabrica extends 182 km and abruptly drops 1,500 m some 30 km from the coast. To the west lie the hills of the northwest region and to the east the Basque mountains that link them to the Pyrenees.

The Sistema Iberico extends from the Cordillera Cantabrica southeastward and, close to the Mediterranean, spreads out from the Rio Ebro to the Rio Jucar. The barren, rugged slopes of this mountain range cover an area of close to 21,000 square kilometers. The mountains exceed 2,000 m in their northern region and reach a maximum height of over 2,300 m east of the headwaters of the Rio Duero. The extremely steep mountain slopes in this range are often cut by deep, narrow gorges.

Other Mountainous Regions

External to the Meseta Central lie the Pyrenees in the northeast and the Sistema Penibetico in the southeast. The Pyrenees, extending from the eastern edge of the Cordillera Cantabrica to the Mediterranean Sea, form a solid barrier and a natural border between Spain and both France and Andorra that, throughout history, has effectively isolated the countries from each other. Passage is easy in the relatively low terrain at the eastern and western extremes of the mountain range; it is here that international railroads and roadways cross the border. In the central section of the Pyrenees, however, passage is difficult. In several places, peaks rise above 3,000 m; the highest, Pico de Aneto, surpasses 3,400 m.

The Sistema Penibetico extends northeast from the southern tip of Spain, running parallel to the coast until it merges with the southern extension of the Sistema Iberico near the Rio Jucar and with the eastern extension of the Sierra Morena. The Sierra Nevada, part of the Sistema Penibetico south of Granada, includes the highest mountain on the peninsula and continental Spain, Mulhacen, which rises to 3,479 m. Other peaks in the range also surpass 3,000 m.

Lowland Regions

The major lowland regions are the Andalusian Plain in the southwest, the Ebro Basin in the northeast, and the coastal plains. The Andalusian Plain is essentially a wide river valley through which the Rio Guadalquivir flows. The river broadens out along its course, reaching its widest point at the Golfo de Cadiz. The Andalusian Plain is bounded on the north by the Sierra Morena and on the south by the Sistema Penibetico; it narrows to an apex in the east where these two mountain chains meet. The Ebro Basin is formed by the Rio Ebro valley, contained by mountains on three sides—the Sistema Iberico to the south and west, the Pyrenees to the north and east, and their coastal extensions paralleling the shore to the east. Minor low-lying river valleys close to the Portuguese border are located on the Tagus and the Rio Guadiana.

The Coastal Plains regions are narrow strips between the coastal mountains and the seas. They are broadest along the Golfo de Cadiz, where the coastal plain adjoins the Andalusian Plain, and along the southern and central eastern coasts. The narrowest coastal plain runs along the Bay of Biscay, where the Cordillera Cantabrica ends close to shore.

The Islands

The remaining regions of Spain are the Balearic and the Canary Islands, the former located in the Mediterranean Sea and the latter in the Atlantic Ocean. The Balearic Islands, encompassing a total area of 5,000 square kilometers, lie 80 kilometers off Spain's central eastern coast. The mountains that rise up above the Mediterranean Sea to form these islands are an extension of the Sistema Penibetico. The archipelago's highest points, which reach 1,400 meters, are in northwestern Mallorca, close to the coast. The central portion of Majorca is a plain, bounded on the east and the southeast by broken hills.

The Canary Islands, ninety kilometers off the west coast of Africa, are of volcanic origin. The large central islands, Tenerife and Gran Canaria, have the highest peaks. Pico de Las Nieves, on Gran Canaria, rises to 1,949 meters, and the Teide, on Tenerife, to 3,718 meters. Teide, a dormant volcano, is the highest peak of Spain and the third largest volcano in the world from its base.

Drainage

Of the roughly 1,800 rivers and streams in Spain, only the Tagus is more than 960 kilometers long; all but 90 extend less than 96 kilometers. These shorter rivers carry small volumes of water on an irregular basis, and they have seasonally dry river beds; however, when they do flow, they often are swift and torrential. Most major rivers rise in the mountains rimming or dissecting the Meseta Central and flow westward across the plateau through Portugal to empty into the Atlantic Ocean. One significant exception is the river with the most abundant flow in Spain, the Ebro, which flows eastward to the Mediterranean. Rivers in the extreme northwest and in the narrow northern coastal plain drain directly into the Atlantic Ocean. The northwestern coastline is also truncated by rias, waterbodies similar to fjords.

The major rivers flowing westward through the Meseta Central include the Duero, the Tagus, the Guadiana, and the Guadalquivir. The Rio Guadalquivir is one of the most significant rivers in Spain because it irrigates a fertile valley, thus creating a rich agricultural area, and because it is navigable inland, making Seville the only inland river port for ocean-going traffic in Spain. The major river in the northwest region is the Mino.

El Atazar Dam is a major dam built near Madrid to provide a water supply.

Climate

Due to both its geographical situation which exposes only its northern part to the Jet Stream's typical path and its orographic conditions, the climate in Spain is extremely diverse.

Peninsular Spain experiences three major climatic types: Continental, Oceanic, and Mediterranean.

Continental Climate

The locally generated tempered continental climate covers the majority of peninsular Spain, influencing the Meseta Central, the adjoining mountains to the east and the south, and the Ebro Basin. A continental climate is characterized by wide diurnal and seasonal variations in temperature and by low, irregular rainfall with high rates of evaporation that leave the land arid. Annual rainfall generally is 300 to 640 mm (11.8 to 25.2 in); most of the Meseta region receives about 500 mm (19.7 in). The northern Meseta, the Sistema Central, and the

Ebro Basin have two rainy seasons, one in spring (April-June) and the other in autumn (October-November), with late spring being the wettest time of the year. In the southern Meseta also, the wet seasons are spring and autumn, but the spring one is earlier (March), and autumn is the wetter season. Even during the wet seasons, rain is irregular and unreliable. Continental winters are cold at "1 °C (30.2 °F), with strong winds and high humidity, despite the low precipitation. Except for mountain areas, the northern foothills of the Sistema Iberico are the coldest area, and frost is common. Summers are warm and cloudless, producing average daytime temperatures that reach 21 °C (69.8 °F) in the northern Meseta and 24 °C (75.2 °F) to 27 °C (80.6 °F) in the southern Meseta; nighttime temperatures range from 7 °C (44.6 °F) to 10 °C (50 °F). The Ebro Basin, at a lower altitude, is extremely hot during the summer, and temperatures can exceed 40 °C (104 °F). Summer humidities are low in the Meseta Central and in the Ebro Basin, except right along the shores of the Rio Ebro, where humidity is high.

Oceanic Climate

An oceanic climate prevails in the northern part of the country, often called "Green Spain", from the Pyrenees to the northwest region, characterized by relatively mild winters, warm but not hot summers, and generally abundant rainfall at 1,000 mm (39.4 in) spread out over the year, with the driest month above 30 mm (1.2 in). Temperatures vary only slightly, both on a diurnal and a seasonal basis, averages range from 9 °C (48.2 °F) in January to 21 °C (69.8 °F) in July. The moderating effects of the sea, however, abate in the inland areas, where temperatures are more extreme than temperatures on the coast.

Distance from the Atlantic Ocean also affects precipitation, and there is less rainfall in the east than in the west. Autumn (October through December) is the wettest season, while July is the driest month. The high humidity and the prevailing off-shore winds make fog and mist common along the northwest coast, this phenomenon is less frequent a short distance inland, however, because the mountains form a barrier keeping out the sea moisture.

Mediterranean Climate

The Mediterranean climate region roughly extends from the Andalusian Plain along the southern and eastern coasts up to the Pyrenees, on the seaward side of the mountain ranges that parallel the coast. Total rainfall in this region is concentrated mostly in late autumn-winter and spring periods. The rain pattern is often irregular which make drought periods likely. Temperatures in the Mediterranean region are higher in winter, and diurnal temperature changes more limited, than those of the continental inland region. Temperatures in January normally average 10 °C (50 °F) to 13 °C (55.4 °F) in most of the Mediterranean region, getting cooler in the northeastern coastal area north of Barcelona. In winter, temperatures inland in the Andalusian Plain are slightly lower than those on the coasts. Temperatures in July and August average 22 °C (71.6 °F) to 27 °C (80.6 °F) on the coast and 29 °C (84.2 °F) to 31 °C (87.8 °F) farther inland, with high humidity. The Mediterranean region is marked by Leveche winds: hot, dry, easterly or southeasterly air currents that originate over North Africa. Episodes of these winds, which sometimes carry fine Saharan dust are more likely in spring, associated to a sudden, usually short-lived, rise of the temperatures. A cooler easterly wind, the Levante, funnels between the Sistema Penibetico and the Atlas Mountains of North Africa. These easterly winds are the ones which most influence the Mediterranean climate, for they are mild in temperature and humid.

Other Climate Types

Besides the above three major types, there are important exceptions, as follows:

- A Semiarid climate in the Southeasternmost part of Spain (covering most of Alicante, Murcia and Almeria provinces). Summers here are hot to very hot and winters mild to cool. Very dry, virtually sub-desertic, rainfall as low as 150 mm (5.9 in) a year in the Cabo de Gata which is reported to be the driest place in Europe.
- The highest section of the Pyrenees and Sierra Nevada qualify as Alpine climate.

- The Canary Islands are a Subtropical climate in terms of temperature, being these mild and stable (18 °C/ 64.4 °F to 24 °C/75.2 °F) throughout the year. In terms of precipitation, the Eastern islands are semiarid and moister the westernmost ones, with some very wet areas in the mountains of Gomera and La Palma, the cloud forest known as laurisilva. Then, the southern Mediterranean coast (Malaga and Granada's coastal strip) shares traits with both Mediterranean and Subtropical.

Extreme Temperature Records in Selected Cities

The record of temperatures in Spain have been,-32°C in Lago Estangento in the Pyrenees (Lerida) for the coldest, for the hottest 47°C in Seville and Murcia.

Population Geography

Largest Cities by Population

- Madrid 3,228,359
- Barcelona 1,582,738
- Valencia 797,654
- Seville 709,975
- Zaragoza 626,081
- Malaga 547,105
- Murcia 391,146
- Las Palmas de Gran Canaria 377,600
- Palma 367,277
- Bilbao 353,567
- Valladolid 321,143
- Cordoba 318,628
- Alicante 305,911
- Vigo 292,566
- Gijon 270,875
- Hospitalet de Llobregat 246,415

• A Coruna (Corunna)	243,902
• Granada	237,663
• Vitoria-Gasteiz	223,257
• Santa Cruz de Tenerife	220,022
• Badalona	214,440
• Oviedo	207,699
• Elche	207,163
• Mostoles	201,789
• Terrassa	200,000
• Pamplona	198,750

Biggest Metropolitan Areas

The most important metropolitan areas and their 2007 populations are:

1.	Madrid	5,603,285
2.	Barcelona	4,667,136
3.	Valencia	1,671,189
4.	Sevilla	1,294,081
5.	Bilbao	950,829
6.	Malaga	897,563
7.	Asturias (Gijon-Oviedo)	857,079
8.	Alicante-Elche	748,565
9.	Zaragoza	731,803
10.	Vigo	662,412
11.	Las Palmas de Gran Canaria	616,903
12.	Bahia de Cadiz (Cadiz-Jerez de la Frontera)	615,494
13.	Santa Cruz de Tenerife	573,825
14.	Murcia	563,272
15.	Palma de Mallorca	474,035
16.	Granada	472,638
17.	San Sebastian	402,168
18.	Tarragona	406,042

19.	A Coruna	403,007
20.	Valladolid	400,400
21.	Santander-Torrelavega	391,480
22.	Cordoba	323,600
23.	Pamplona	309,631

Islands

Islander population:

1.	Tenerife	886.033
2.	Mallorca	846.210
3.	Gran Canaria	829.597
4.	Lanzarote	132.366
5.	Ibiza	113.908
6.	Fuerteventura	94.386
7.	Menorca	86.697
8.	La Palma	85.933
9.	La Gomera	22.259
10.	El Hierro	10.558
11.	Formentera	7.957
12.	Arosa	4.889
13.	La Graciosa	658
14.	Tabarca	105
15.	Ons	61

Resources and Land Use

Natural resources: coal, lignite, iron ore, uranium, mercury, pyrites, fluorspar, gypsum, zinc, lead, tungsten, copper, kaolin, potash, hydropower, arable land.

Land use:

- *Arable land:* 30%
- *Permanent crops:* 9%
- *Permanent pastures:* 21%
- *Forests and woodland:* 32%
- *Other:* 8% (1993 est.)

Irrigated land: 34,530 km² (1993 est.)

Environmental Concerns

Natural hazards: periodic droughts, wildfires

Environment-Current Issues:

- pollution of the Mediterranean Sea from raw sewage and effluents from the offshore production of oil and gas
- water quality and quantity nationwide
- air pollution
- deforestation
- desertification.

Environment-International Agreements:

- *Party to:* Air Pollution, Air Pollution-Nitrogen Oxides, Air Pollution-Sulphur 94, Air Pollution-Volatile Organic Compounds, Antarctic-Environmental Protocol, Antarctic Treaty, Biodiversity, Climate Change, Endangered Species, Environmental Modification, Hazardous Wastes, Law of the Sea, Marine Dumping, Marine Life Conservation, Nuclear Test Ban, Ozone Layer Protection, Ship Pollution, Tropical Timber 83, Tropical Timber 94, Wetlands, Whaling
- *Signed, but not ratified:* Air Pollution-Persistent Organic Pollutants, Climate Change-Kyoto Protocol, Desertification

Maritime claims

- *contiguous zone:* 24 nmi (44.4 km; 27.6 mi)
- *exclusive economic zone:* 200 nmi (370.4 km; 230.2 mi) (applies only to the Atlantic Ocean)
- *territorial sea:* 12 nmi (22.2 km; 13.8 mi)

Geography of Portugal

Portugal is a coastal nation in southwestern Europe. Its mainland is located on the Iberian Peninsula and borders only

Spain, to the east and north, with a total of 1,214 km (754 mi) of borders. Despite this total of km for the border with Spain, Portugal does not recognise the border from the delta of the River Caia to the delta of the River Cuncos since 1801's occupation of the Olivenca territory by the Spanish Kingdom. This territory, though under Spanish occupation remains part of Portugal and no border is henceforth recognised in this area. It is bordered on the west with a 1,793 km (1,114 mi) coastline along the North Atlantic Ocean. Portugal also controls the Atlantic islands of Azores and Madeira Islands, which are strategic locations along western sea approaches to the Strait of Gibraltar between the Atlantic and Mediterranean Sea. In total, the country has an area of 92,090 km^2 (35,560 sq mi) of which 91,470 km^2 (35,320 sq mi) is land and 620 km^2 (240 sq mi) water.

Statistics

Maritime claims:

Continental shelf: 200 m depth or to the depth of exploitation.

Exclusive economic zone: 200 nmi (370.4 km; 230.2 mi).

Territorial sea: 12 nmi (22.2 km; 13.8 mi).

Natural hazards: The Azores are occasionally subject to very strong earthquakes, as is the continental coast. Wildfires occur mostly in the summer in mainland Portugal and extreme weather in the form of strong winds and floods also occurs mainly in winter. The Azores are occasionally stricken by tropical cyclones such as Hurricane Jeanne (1998) and Hurricane Gordon (2006).

The Continent

Continental Portugal is split in two by its main river, the Tagus (*Tejo*). To the north the landscape is mountainous in the interior areas with plateaus, cut by four breakings lines that allow the development of more fertile agricultural areas.

The south down as far as the Algarve features mostly rolling plains with a climate somewhat warmer and drier than the cooler and rainier north. Other major rivers include the

Douro, the Minho and the Guadiana, similar to the Tagus in that all originate in Spain. Another important river, the Mondego, originates in the Serra da Estrela (the highest mountains in mainland Portugal at 1,993 m).

No large natural lakes exist in Continental Portugal, where the largest inland water surfaces are dam-originated reservoirs, (such as the Alqueva reservoir with 250 km^2, the largest in Europe). However, there are several lagoons in Portugal with a river as origin:

- in Serra da Estrela, the Comprida Lagoon (*Lagoa Comprida*) and the Escura Lagoon (*Lagoa Escura*) were formed from ancient glaciers.
- the Pateira de Fermentelos is a small natural lagoon near Aveiro

Some lagoons are near the Atlantic Ocean. For instance, the Albufeira Lagoon (in the Setubal Peninsula) and Obidos Lagoon (near Foz do Arelho, Obidos). In the Azores archipelago lagoons were formed in extinct volcanos. Lagoa do Fogo and Lagoa das Sete Cidades (two small lakes connected by a narrow passage) are the most famous lakes in Sao Miguel Island.

The Islands

In addition to continental Europe, Portugal consists of two Autonomous Regions in the Atlantic Ocean, consisting of the archipelagos of Madeira and Azores. Madeira is located on the African Tectonic Plate, and comprises the main island of Madeira, Porto Santo and the smaller Savage Islands. The Azores, which are located between the junction of the African, European and North American Plates, straddle the Mid-Atlantic Ridge.

There are nine islands in this archipelago, usually divided into three groups (Western, Central and Eastern) and several smaller Formigas (rock outcroppings) located between Sao Miguel and Santa Maria Islands. Both island groups are volcanic in nature, with historic volcanology and seismic activity persisting to the present time. In addition, there are several submarine volcanos in the Azores (such as Dom Joao de Castro Bank), that have erupted historically (such as the Serrata eruption off the coast of Terceira Island). The last major volcanic event occurred in 1957-58 along the western coast of Faial Island, which formed the Capelinhos Volcano. Seismic events are common in the Azores.

The Coast

The Portuguese coast is extensive, it has 943 km (586 mi) in continental Portugal, 667 km in the Azores, 250 km in Madeira and in the Savage Islands. The Portuguese coast developed fine beaches, the Algarve ones are worldwide famous. In Porto Santo Island, a dune formation is appealing to many tourists. An important feature in its coast is the Ria de Aveiro (near Aveiro, a town called "The Portuguese Venice"), a delta with 45 km in length and a maximum of 11 km width, rich in fish and sea birds.

There are four main channels, between them several islands and islets, and it is where Vouga, Antua, Boco, and Fontao rivers meet the ocean. Since the 16th century, the formation of a sort of narrow headlands formed a lagoon, seen as one of the most remarkable hydrographic features of the Portuguese coast.

Due to these characteristics, the region is an ancient producer of a famous salt, and during the Roman Empire it exported its salt to Rome, seen as a precious product. Another interesting feature of the Portuguese coast is Ria Formosa, located in the Algarve, with some sandy islands and a mild and pleasant climate characterized by warm but not very hot summers and generally mild winters.

Exclusive Economic Zone

Portugal has the 3rd largest Exclusive Economic Zone of the EU and the 11th in the world. The seazone over which the Portuguese have special rights over the exploration and use of marine resources, has 1,727,408 km^2.

- Continental Portugal 327,667 km^2
- Azores Islands 953,633 km^2
- Madeira Islands 446,108 km^2
- Total: 1,727,408 km^2.

Climate

Portugal has a Mediterranean climate, *Csa* in the south and *Csb* in the north, according to the Koppen climate classification. The annual average temperature in mainland Portugal varies from 12–13 °C (53.6–55.4 °F) in the mountainous interior north to 17–18 °C (62.6–64.4 °F) in the south (in general the south is warmer and drier than the north). The Madeira and Azores archipelagos have a narrower temperature range with the annual average temperature sometimes exceeding 20 °C (68 °F) in the south coast of Madeira Island.

Extreme temperatures occur in the mountains in the northeast of the country in winter, where they may fall to "15 °C (5.0 °F), and in southeastern parts in the summer, sometimes exceeding 45 °C (113 °F). The official absolute extreme temperatures are "16 °C (3.2 °F) in Penhas da Saude on 4 February 1954 and 47.4 °C (117.3 °F) in Amareleja in the Alentejo region, on 1 August 2003. There are, however, unofficial records of 50.5 °C (122.9 °F) on 4 August 1881 in Riodades, Sao Joao da Pesqueira. It's very plausible since this region is known for its microclimate. The annual average rainfall varies from a bit more than 3,000 mm (118.1 in) in the mountains in the north to less than 600 mm (23.6 in) in southern parts of Alentejo.

The country has around 2500-3200 hours of sunshine a year, an average of 4-6h in winter and 10-12h in the summer, with higher values in the southeast and lower in the northwest. The sea surface temperature is higher in the south coast where it varies from 15–16 °C (59–60.8 °F) in January to 22–23 °C (71.6–73.4 °F) in August, occasionally reaching 25 °C (77 °F); on the west coast the sea surface temperature is around 14–15 °C (57.2–59 °F) in winter and 18–20 °C (64–68 °F) in the summer, with higher values as one goes southwards.

Both the Azores and the Madeira Islands have a subtropical climate with mild, rainy winters and mainly dry, warm summers. The Savage Islands, that belong to the Madeira archipelago, have an arid climate with an annual average rainfall of around 150 mm (5.9 in). The sea surface temperatures in these archipelagos vary from 16–18 °C (60.8–64.4 °F) in winter to 23–24 °C (73.4–75.2 °F) in the summer, occasionally reaching 26 °C (78.8 °F).

Environment

Environment-current issues: soil erosion; air pollution caused by industrial and vehicle emissions; water pollution, especially in coastal areas.

Environment-international agreements:

Party to: Air Pollution, Biodiversity, Climate Change, Desertification, Endangered Species, Hazardous Wastes, Law of the Sea, Marine Dumping, Marine Life Conservation, Ozone Layer Protection, Ship Pollution, Tropical Timber 83, Tropical Timber 94, Wetlands

Signed, but not ratified: Air Pollution-Persistent Organic Pollutants, Air Pollution-Volatile Organic Compounds, Climate Change-Kyoto Protocol, Environmental Modification, Nuclear Test Ban

Terrain: mountainous and hilly north of the Tagus River, rolling plains in south

Elevation extremes:

Lowest point: Atlantic Ocean 0 m

Highest point: Ponta do Pico (Pico or Pico Alto) on Ilha do Pico in the Azores 2,351 m

Natural resources: fish, forests (cork), tungsten, iron ore, uranium ore, marble, arable land, hydroelectric power

Land use:

Arable land:	26%
Pe**r**manent	0%
Rops:	9%
Permanent pastures:	9%
Forests and woodland:	36%
Other:	20% (1993 est.)

Irrigated land: 6,300 km² (1993 est.)

Overall Characteristics

The Iberian peninsula extends from the southernmost extremity at Punta de Tarifa to the northernmost extremity at

Estaca de Bares Point over a distance between lines of latitude of about 865 kilometres (537 mi) based on a degree length of 111 km per degree, and from the westernmost extremity at Cabo da Roca to the easternmost extremity at Cap de Creus over a distance between lines of longitude at 40° N latitude of about 1,155 kilometres (718 mi) based on an estimated degree length of about 90 km for that latitude. The irregular, roughly octagonal shape of the peninsula contained within this spherical quadrangle was compared to an ox-hide by the geographer, Strabo.

Approximately 3/4 of the octagon is the Meseta Central, a low and rolling plateau of up to several hundred meters in altitude. It is located roughly in the center, staggered slightly to the east and tilted slightly toward the west. (The conventional center of the Iberian Peninsula has long been considered to be Getafe just south of Madrid.) It is ringed by mountains and contains the sources of most of the rivers, which find their way through gaps in the mountain barriers on all sides.

Coastline

The coastline of the Iberian Peninsula is 3,313 km (2,059 mi), 1,660 km (1,030 mi) on the Mediterranean side and 1,653 km (1,027 mi) on the Atlantic side. The coast is a drowned one, with sea levels having risen from a minimum of 115 metres (377 ft) to 120 metres (390 ft) lower than today at the Last Glacial Maximum (LGM) to its current level at 4000 years BP. The coastal shelf created by sedimentation during that time remains below the surface; however, it was never very extensive on the Atlantic side, as the continental shelf drops rather steeply into the depths. An estimated 700 km (430 mi) length of Atlantic shelf is only 10 km (6.2 mi) to 65 km (40 mi) wide. At the 500 m (1,600 ft) isobath, on the edge, the shelf drops off to 1,000 m (3,300 ft).

The submarine topography of the coastal waters of the Iberian Peninsula has been studied extensively in the process of drilling for oil. Ultimately the shelf drops into the Bay of Biscay on the north (an abyss), the Iberia abyssal plain at 4,800 m (15,700 ft) on the west and Tagus abyssal plain to the south. In the north between the continental shelf and the abyss is an extension, the Galicia Bank, a plateau containing also the Porto, Vigo and Vasco da Gama seamounts, creating the Galicia interior basin. The southern border of these features is marked by Nazare Canyon, splitting the continental shelf and leading directly into the abyss.

Mountains

Mountains consist mainly of serrated ridges aligned in an east-west direction, due to the orogenic factors of the region's geologic history. Rivers generally flow through the valleys between the ridges. In a counterclockwise direction, the major mountain ranges are: the Pyrenees crossing the isthmus of the peninsula so completely as to allow no passage except by mountain road or trail or coastal road, the Cantabrian Mountains perched on the northern coastline, a series of ridges straddling Portugal and Spain: the Sierra de Guadarrama, the Sierra de Gredos, the Sierra de Gata, and the Serra da Estrela; across the south: the Sierra Morena and the Sierra Nevada.

Ecology

The woodlands of the Iberian Peninsula are distinct ecosystems on the Iberian Peninsula, spanning mainly Spain and Portugal. Although the various regions are each characterized by distinct vegetation, there are some similarities across the peninsula.

While the borders between these regions are not clearly defined, there is a mutual influence which makes it very hard to establish boundaries and some species find their optimal habitat in the intermediate areas.

Origin and Characteristics

The flora of the peninsula, because of bio-historical, geographical, geological, and orographic conditions, are among of the richest and most varied of all European flora, rivaled only by such countries as Greece and Italy; it is estimated that the Iberian Peninsula has more than 8,000 distinct species of plants, many of them endemic. With more than 8,500 plant species, Spain has the richest flora in Europe. It is now known that the Mediterranean Sea went through great changes of climate and vegetation, among them changes in sea level and variations in the relative positions of the continental plates of Europe and Africa.

The Iberian Peninsula, located on an important route between Africa and Europe, was enriched by the arrival, following the climate change, of wetland plants, thermophilic plants (those that require a great deal of heat), xerophilic plants (those that require a dry climate), orophilic (sub-alpine) plants, Boreo-alpine plants, and so on, many of which managed to remain, thanks to the diversity of environments that exist in the mountain ranges, and which allowed them to rise in altitude if the climate was too warm, or descend if it became too cold. The geological complexity of the majority of Iberian mountains, especially of the Cordillera Betica, Sistema Iberico, and Pyrenees, also greatly increased the number of new environments to which it was possible to adapt, resulting in today's wide variety of flora.

The Eurosiberian Region

The "Eurosiberian" Atlantic zone extends through northern Portugal, Galicia, Asturias, Cantabria, the Basque Country and the western and central Pyrenees. It is characterized by a humid climate which is moderated by the influence of the ocean, with somewhat cold winters and the lack of a distinct dry season. The mainland extends to the north of Portugal, the greater part of Galicia, Asturias, Cantabria, the Basque Country, northwest of Navarre, and western Pyrenees. However, its influence in the form of communities or defined species extends inwards, especially in the north and west.

The vegetation is deciduous oak forest: both Sessile Oaks (*Quercus petraea*) and English Oaks (*Quercus robur*), with European Ash (*Fraxinus excelsior*) and hazels in the coolest and deepest soil at the bottom of the valley. The mountain layer is characterised by the presence of beeches and at times, in the Pyrenees, by Silver Firs (*Abies alba*); these beeches and silver firs occupy the cool slopes with shallow scil. The Mediterranean influence is felt in the presence of Holm Oaks (*Quercus ilex*) with laurel, which are situated on the warmest crests and slopes, especially above chalky soil, where the dryness becomes more pronounced. Improvement by humans has transformed much of this woodland into meadows, which conserve at their edges remnants of the *setos* of the species of the primitive forest. The natural edge is formed by *setos* and clumps of thorny shrubs growing in glades and clearings, such as the wild rose, blackberry bushes, blackthorn, hawthorn and other more or less thorny shrubs; this role can also be filled by *los piornales* and clumps of broom. The major forests in this area are beech, oak, birch, and fir.

Beech Forests

Beech forests are found in the mountain layer of the Iberian Eurosiberian region from 800 to 1500 metres up. The soil is cool, as often chalky as siliceous (rich in silica), and nearly always acidified by rain. The layer is characterised by the beech tree. The beech tree projects a deep shadow, and so its dense foliage usually excludes other woodland species. It therefore has little undergrowth.

In spite of their Atlantic character, these forests reach Moncayo, in the centre of the peninsula. The southernmost are at the Hayedo de Montejo (in the autonomous community of Madrid) and in the northernmost area of the province of Guadalajara, in the Parque Natural del Hayedo de Tejera Negra (Cantalojas), and Somosierra-Ayllón. The forests seek watercourses and shade, and so their reforestation is very difficult and they are being displaced by the Pyrenean Oak (*Quercus pyrenaica*). The Irati "rainforest", of some 170 square kilometres in the Navarran Pyrenees, is one of the most important beech and fir forests in Europe.

Oak Forests

Oak forests, above all of English Oak (*Quercus robur*), are the most common in the Atlantic zone. They represent the typical forest floor formation of basal trees, extending to an altitude of some 600 metres. In higher regions, as one ascends the mountains, they yield to beech forests; at the bottom of the valleys they are replaced by ash trees and hazel tree groves. There are two main types of oak: the English Oak and the Sessile Oak (*Quercus petraea*). The latter extends furthest into the interior and highest in altitude, but plays a secondary part; in general, when the climate begins to show its continental character, these oak forests are replaced by Pyrenean oak.

The land on which these oaks stood is the most altered, as it is well suited to meadows and crops. Oaks are often accompanied by chestnut trees and birches. When these forests degrade, they are taken over by thorny plants, *piornales,* and in the final extreme heather and gorse. The English Oak would have been indigenous to a large part of the area currently occupied by pine forests and eucalyptus.

Birch Forests

Along the Atlantic coast, birches (*Betula* species) form small enclaves or copses at the foot of rocky cliff edges or in the clearings of beech forests, on poor or acidic soils, accompanied by Aspen (*Populus tremula*) and Mountain Ash (*Sorbus aucuparia*). Birch may also grow in pure stands near the beech forests, in the mountainous areas on siliceous bedrock; these areas are typically of only small extent and generally quite patchy with Sessile Oak (*Quercus petraea*) and trees of the genus *Sorbus*.

Fir Forests

The Silver Fir (*Abies alba*) is found on the cool, deep-soiled slopes of the flanks of the Pyrenees, from Navarre to Montseny, forming pure fir woods or, more often, mixed forests with beech trees. The most important areas are in Lleida (Lerida), with 170 square kilometres. In altitude, it extends from 700 to 1700 meters, but its main areas are localized in more humid and dark valleys; these forests are dark, with acidic soil, due to the decomposition of the needles of evergreens. At higher altitudes, they are often replaced by Black Pine (*Pinus uncinata*). These spruce forests sometimes contain Maples (*Acer pseudoplatanus*) and their undergrowth is very similar to that of the beech forest. Like these, they are distinctly Euro-Iberian.

The Mediterranean Region

The Mediterranean region occupies the rest of the peninsula (the greater part of it), as well as the Balearic Islands. The principal characteristic of the region is the existence of a quite lengthy period of summer drought, which may last anywhere from 2 to 4 months, but which, regardless of length, is always quite distinct. Rainfall can range from 1500 mm to less than 350 mm. Temperatures range from regions that have no frost for many years to those that reach-20 °C, or even lower, every winter.

If one ignores, for the moment, the influence of the mountains, the typical Mediterranean peninsular forest is made up of evergreen trees: oak forests, cork oaks, wild olives, juniper, and so on. These are accompanied or replaced in the warmer regions and eroded by forests of Aleppo Pine (*Pinus halepensis*) and in areas of sandy ground and fixed sand dunes by juniper and Stone Pine forests.

Exceptions to the rule are the more arid region in the southeast, the lower regions of the provinces of Murcia and Almeria, where the only vegetation is the European Fan Palm (*Chamaerops humilis*), and thorny thickets of blackthorn and at higher altitudes, Kermes Oak groves and mastic (*Pistacia lentiscus*). The same can be said of the salty or endorheic zones, with great differences in temperature, such as the depression of the Ebro, Hoya de Baza and the chalky marls further inland.

Pyrenean Oaks

Of all the oaks, the Pyrenean Oak (*Quercus pyrenaica*) is the most resistant to drought and the continental-style climate. These forests, of a sub-Atlantic character, often represent the shift from Mediterranean vegetation to Atlantic vegetation. They cover a wide area of the peninsula and are of great importance, above all on the mountain ranges in the centre of the peninsula; from the interior of Galicia and extending south of the Cordillera Cantabrica they extend throughout the Sistema Central, reaching, to the south, (though scarce by the time they reach this region) the Sierra Nevada and Cadiz.

They usually extend from some 700 to 800 metres to some 1500 to 1600 m in altitude. They prefer siliceous soil and, as altitude increases, they replace the damp oak forests and Cork

Oak; on the high ground they give way to Scots Pine (*Pinus sylvestris*) forests or to *los piornales serranos* with creeping juniper. In areas where the influence of the Atlantic is more evident, they are taken over by heather and *Erica australis*; in the remainder, in clearings and in more degraded phases, rock rose (*Cistaceae*) mixed with laurel leaf and Spanish Lavender (*Lavandula stoechas*) is more frequent. Their natural range is usually covered by forests of Scots Pine or Maritime Pine.

Groves, Riparian Forests and Valley Floors

In the groves, riparian forests, and valley floors, there are enclaves of deciduous forests that favour for the moist ground, which is constant almost all year; this allows them to avoid the consequences of the summer drought so characteristic of a Mediterranean climate.

There we see a characteristic pattern, as we move outward from the edge of the riverbed, so that the woodlands that are most dependent on the aquifer are on the riverbank, that is, (alders and willow groves) and those less dependent on water are located further away, such as the (ash, elm, and poplar groves).

These woodlands are formed by willows, poplars, alder, ash, elm and sometimes by Pyrenean Oak (*Quercus pyrenaica*), lime trees, birches and hazel trees. When the humidity starts to diminish in barren areas of the Ebro valley, the Levante and the southern half of the peninsula, the dryness is often accompanied by an increase of salts in the ground; under such conditions one can find formations of tamarisk shrubs, oleanders and giant reeds (*Saccharum ravenae*), sometimes accompanied by heather. In soils rich in silica but not in salts, such as those of the Sierra Morena and the Montes de Toledo, spurge appears, accompanied, in the warmest places, by oleanders and tamarisks.

In the lowlands further inland, above all in the marls and clay soils, Field Elms (*Ulmus minor*) and poplar groves are more common, with occasional ashes and willows. At the bottom of granite valleys and on the siliceous riverbanks, there are very typical formations of ash with Pyrenean Oak, especially at the foot of thin interior mountain ranges. The sheltered gorges of the Serrania de Cuenca have mixed riparian forests of lime and hazel trees, with ashes, willows and Wych Elm (*Ulmus glabra*). Because these woods occupied some of the most fertile lands, where people have planted orchards since ancient times, they have not been well preserved.

Spanish Firs

The Spanish Fir (*Abies pinsapo*) is a true relic which has remained preserved in some mountain ranges around Malaga and Cadiz. Spanish firs are related to the North African spruce forests of the Yebala range, in Morocco. They come into contact with Algerian Oak (*Quercus canariensis*) and other oak trees and sometimes even form mixed communities with these. Among the woody species also found in these forests are the hawthorn, barberry, Butcher's broom (*Ruscus aculeatus*), *Viburnum tinus*, ivy, and *Daphne laureola*.

It forms dense and dark forests in very distinct enclaves, in areas with high rainfall (from 2,000 to 3,000 mm, due to the sudden cooling, with elevation, of humid winds), at altitudes of over 1,000 metres. The forest has abundant moss and lichens, but very few shrubs and herbaceous plants. In all cases, Spanish fir occupies high mountain zones (such as the Sierra de las Nieves, Sierra Bermeja, and Sierra de Grazalema).

Holm Oak Forests

Forests of Holm Oak (*Quercus ilex*) form natural forests in most of the Mediterranean region as well as penetrating into the warmer sun-exposed areas and hillsides of the Atlantic region; they extend from sea level, with the subspecies *ilex*, to an altitude of 1400 metres, in some mountains and high plains of the interior; in the continental zone, the oak found is the subspecies *rotundifolia*, more resistant to such a climate.

The holm oak can also be found at higher altitudes, but as isolated trees, not forming forests. Coastal oak forests and those of sublittoral mountains are extraordinarily rich and varied, with a variety of shrubs and lianas; often accompanied by bramble, honeysuckle, ivy, *Viburnum tinus*, Butcher's Broom and, in the southwest of the peninsula, wild olive trees. The oak forests of the Balearic Islands are also rich, and incorporate characteristic species of the islands, such as the Balearic Cyclamen (*Cyclamen balearicum Willk.*).

Towards the interior of the peninsula, these forests become progressively more scarce: as the continental characteristics of the climate become stronger, the species most sensitive to cold become steadily more scarce.

The continental groves, on soils lacking lime (calcium oxide), tend to be rich in junipers (*Juniperus oycedrus*) and are superseded at higher altitudes and on cooler slopes by Pyrenean Oaks. This phenomenon is apparent in the Sierra de Guadarrama: when the oak forests have been destroyed, the soil is so poor and the environmental conditions so unfavourable, that it leads to ragged thickets dominated by common rock rose, Spanish lavender and rosemary. On limy soils something similar takes place, above all at altitudes of over 900 metres, oaks are accompanied by Spanish Juniper (*Juniperus thurifera*) and the scarcity of shrubs is such that the same Holm Oak (*Q. ilex* subsp. *rotundifolia* or *ballota*) dominates almost entirely on its own the first phases of deterioration of the forest. The degradation caused by burning or felling leads to thickets of Scorpion's thorn (*Genista scorpius*), thyme and common lavender (*Lavandula angustifolia*).

Cork Oak Forests

Cork oak forests occupy around 10,000 square kilometres on the peninsula, more than half of the world's extension of this kind of forest.

The Cork Oak forest needs siliceous soils of a sandy texture, and a mild, slightly damp climate. Under such conditions it totally or partially displaces the Holm oak; Holm oak may be found in stands of Cork Oak with a certain frequency, as well as Portuguese Oaks (*Quercus faginea* subsp *broteroi*). The area occupied by oak forests corresponds to Catalonia, Menorca and above all to the southwest quadrant of the peninsula. They often alternate with the oaks, which occupy the drier slopes and with the *quejigares* of Algerian oak (*Quercus canariensis*), that occupy the ravines and cool, shady northward slopes.

The cork oak forests often contain wild olives, and like some of the cool groves, are often accompanied by Strawberry trees (*Arbutus unedo*) with Mock Privet (*Phillyrea angustifolia*) which grow in the clearings of these forests and dominate their regressive phases. In western Andalusia, other common components of the ecosystem are the areas of common broom, genus *Teline*.

Quejigares

The term *quejigar* refers to forests of many different characteristics. Forests of Algerian oaks (*Quercus canariensis*) are well-represented in western Andalusia and very patchily by hybridizations in Catalonia and the Cordillera Mariánica. They are the most demanding as to temperature and humidity, and therefore do not usually stray too far from areas with a maritime climate; they prefer the cooler, shady northward slopes, damp meadows and the banks of streams of the lower ground.

In general, they alternate with the Cork Oaks, which they displace in the coolest zones; both prefer siliceous soils. In the clearings and degraded stages of these forests *los piornos* (*Teline sp., Cytisus baeticus*), heather (*Erica arborea, Erica scoparia*), and rock rose (*Halimium lasianthum*) are common.

Los quejigares of Portuguese oaks (*Quercus faginea* subsp *faginea*) are the most typical and common of the peninsula, since they are found from the *Serrania de Ronda* in Andalusia to the lower slopes of the Pyrenees. They are much more resistant to the cold and damp than *Q. Canariensis*; on the other hand, they need cooler and deeper soils than the Holm oaks with which they come into contact.

Although they can grow in any type of soil, in siliceous soils the usually play a secondary part with respect to Holm Oaks, Cork Oaks, and Pyrenean Oaks; it is only on chalky soils that it forms forests of any consideration, especially in the northeast quadrant and centre of the Peninsula. The natural area corresponding to the *quejigo* is frequently the Black pine (*Pinus nigra* subsp *salzmannii*), which often has been extended at its expense.

Los quejigares can often include *maples, serbales,* European serviceberry or snowy mespilus (*Amelanchier ovalis*), common privet (*Ligustrum vulgare L.*) and Common Dogwood (*Cornus sanguinea*); their degradation can lead to extensive thickets of box.

The last *quejigo*, *Quercus faginea* subsp *broteri* requires the most moisture and is least resistant to cold. It is found principally in the southwest quadrant and prefers siliceous soils, somewhat cool. More frequently than in pure concentrations, it is found mixed with Cork Oaks and Holm Oaks.

Pine Forests

The most characteristic natural pine forests are those of *pino negro* (*Pinus uncinata*) and Scots Pine (*Pinus sylvestris*). The former is often associated with *Rhododendron ferrugineum*, blueberries, *Salix pyrenaica* and other shrubby species on the subalpine slopes of the Pyrenees. Over less washed limestone soils it is usually accompanied by Savin Juniper (*Juniperus sabina L.*), Common Juniper (*Juniperus communis* subsp. *hemisphaerica*), and Common Bearberry (*[Arctostaphylos uva-ursi]*). Such forests make up the tree line in most of the Pyrenees, reaching 2400 metres.

The Scots Pine plays the same part in the other peninsular mountains, both siliceous and limy. It is accompanied and superseded at high altitude by *piornales,* dwarf junipers, and hummocky high mountain thickets. Their lower altitudinal limit remains patchy, having been extended at the expense of deciduous forests.

The Maritime pine (*Pinus pinaster*) is found at an intermediate altitude and on generally siliceous soil, which in Galicia goes down to sea level and inland alternates with Pyrenean Oak. Over limestone, the Black Pine (*Pinus nigra* subsp *salzmannii*) plays an important role in many of the mountain ranges in the centre, east and south of the Peninsula;

in limy soil, and at the same altitude, it usually displaces the former. Both are displaced at higher altitudes by the Scots pine.

The warmest of all the pine forests are those of Aleppo Pine (*Pinus halepensis*), which are situated on rocky crests and sunny hillsides. Aleppo Pine is the typical pine of the Mediterranean coast, from sea level to an altitude of 800–1000 metres in the interior; these prefer limy soils.

The Stone Pine (*Pinus pinea*), possibly the most characteristic of all, occupies sandy soils. It grows extensively both in the sandy areas of the lowlands in the provinces of Cadiz and Huelva, as well as points further inland (Valladolid, Cuenca, and Madrid). Finally, especial mention is necessary of the Monterey Pine (*Pinus radiata*), because of its importance in reforestation and managed forests.

Juniper Groves

The juniper groves of Spanish Juniper (*Juniperus thurifera*) constitute a curious formation that occupies the high heaths and *mesetas* of the interior, nearly always above 900 metres in altitude. The principal woodlands of this type are in the Serrania de Cuenca, Sistema Iberico, Alcarria, Maestrazgo and other mountains of the interior.

They do not usually form dense forests, but rather parkland or small woods in meadows. They prefer soils developed over limestone, especially those of an ochre or reddish colour and rich in clay, *de caracter relicto (Terra rosa, Terra fusca);* on occasions, as in the region of Tamajon (Guadalajara), they also colonize siliceous terrain.

They are adapted to an exceptionally harsh continental climate, where practically no other species of tree is a rival; except for the Holm Oak, which occupies some of the old, deforested juniper groves, and the European Black Pine (*Pinus nigra*) which can accompany it with a certain frequency. The Common Juniper (*Juniperus communis* subsp. *hemisphaerica*) is habitually a secondary species of these groves. At high altitudes, they come in contact with forests of Scots pine and with the Savin Juniper; the latter sometimes forming part of the shrubby stratum. The fact that these are to be found mainly in areas which have been exposed for a large part of the Tertiary and over soils considered *relictos*, supposes a great antiquity for such groves.

The harsh climatological conditions, with the surface of the ground undergoing processes of alternate freezing and thawing (cryoturbation), makes difficult the development of elevated brush. In their regressive stages, they tend toward hummocky thickets of *cambrones* (*Genista pumila*) or tomillares *y prados de diente dominados por* dwarf shrubs and Dog's Tooth Grass. At lower altitudes, these groves can also alternate with *espliego y aliaga.*

The Phoenicean Juniper (*Juniperus phoenicea*) habitually plays a secondary part and does not often form dense woods. Only on some rocky shelves or in special environments such as fixed dunes and sandy areas near the coast do they manage to form woodlands of any importance.

Thickets of the High Mediterranean Mountains

The high Mediterranean mountains above 1700 metres presents some special characteristics. Winters are very harsh and long; the thickness of the snow and strong frosts prevent almost any type of biological activity. Once the snow has disappeared, the soil dries quickly due to the strong sun and high temperatures reached in summer. The period of time suitable for the growth of vegetation is therefore very short and for the above-mentioned reasons, the land is dry most of the time. Under such conditions, the forest enters a state of crisis, being replaced by *piornales* (ciste and broom formations) and thickets *pluvinulares* accompanied at lower levels by Scots Pine, often isolated individuals twisted and deformed by the snow.

Siliceous mountains such as the Sistema Central, Serra da Estrela, the Sistema Iberico of the region of Soria, and parts of the Cantabrian mountains, are covered by thickets of *Cytisus purgans* (variously known as "Andorra broom", "Provence broom", or "Spanish gold hardy broom") or Alpine juniper (*Juniperus communis* subsp *alpina*). In the Sierra Nevada, on the other hand, under similar conditions, the *Genista baetica*, is more dominant, sometimes accompanied by *Cytisus purgans* and another species of juniper (*J. communis* subsp. *hemisphaerica*).

In the limestone mountains such as the Maestrazgo and Serrania de Cuenca, a shrubby formation of Savin Juniper (*Juniperus sabina*) accompanied by Scots Pine is characteristic. In the limestone mountains of Andalusia a detectable role is played by thickets *pluvilunares* and hummocks of common buckthorn (*Rhamnus cathartica*).

Shrubby Borders or Undergrowths

From an ecological point of view, shrubby borders are fundamental in forest ecosystems to guarantee the natural regeneration of the woods, as well as to provide food and refuge for the associated fauna. They are made up of spiny shrubs, depending on the forest and climate, such as gorse, box, thyme, and so on.

Stages of Degradation

It is possible to identify successive stages in the process of degradation of these various forest formations, from an optimum state to the final phase of desertification.

These regressive states, in the case of leafy forests, are the following:

1. Dense forest representative of an optimal natural state, characterized by endemic species, compatible with local biological conditions.
2. *Bosque aclarado,* still with a predominance of the native species, but with abundant representation of a variety of species such as holly, maple, and ash. Frequently, leguminous plants predominate in the scrub of the underbrush.
3. Stage of pine forests. The native species have practically disappeared, as has the associated vegetation. Together with the pines, heliophilic (preferring dry and sunny habitats) and invasive brush start to appear,

almost always based on the families *Cistaceae* and *Ericaceae.*

4. The stratum of trees as such disappears, along with its associated species, replaced progressively by brush representative of a very advanced degradation; there is a high frequency of thorny plants (Scorpion's thorn, blackthorn (*Prunus spinosa*), etc.) and a predominance of Lamiaceae *y compuestas* (Tomillo vulgar, heather, pennyroyal (*Mentha pulegium*), etc.).
5. The ground cover is reduced, not only in the size of plants, but also in the area it occupies; now it forms a hebaceous and discontinuous tapestry, with a general predominance of Dog's Tooth Grass. Woody plants are reduced to some thickets, the bedrock being exposed as a consequence of erosion. This is the typical landscape of the steppe.
6. The final stage of regression is represented by desertified ground.

East Atlantic Flyway

The Iberian Peninsula in an important stopover on the East Atlantic flyway for birds migrating from northern Europe to Africa. For example, Calidis ferruginea rests in the region of Cadiz Bay.

In addition to the birds migrating through, some seven million wading birds from the north spend the winter in the estuaries and wetlands of the Iberian Peninsula, mainly at locations on the Atlantic coast. In Galicia are the Ria de Arousa (a home of Pluvialis squatarola), Ria de Ortigueira, Ria de Corme and Ria de Laxe. In Portugal the Aveiro Lagoon hosts Recurvirostra avosetta, Charadrius hiaticula, Pluvialis squatarola and Calidris minuta.

Ribatejo on the Tagus River supports Recurvirostra arosetta, Pluvialis squatarola, Culidris alpina, Limosa lapponica and Tringa totanus. In the Estuário do Sado are Calidris alpina, Numenius arquata, Pluvialis squatarola and Tringa totanus. The Algarve hosts Calidris canutus, Tringa nebularia and

Arenaria interpres. The Marismas de Guadalquivir region of Andalusia and the Salinas de Cadiz are especially rich in wintering wading birds: Charadrius alexandrinus, Charadrius hiaticula, Calidris alba, and Limosa limosa in addition to the others. And finally, the Ebro delta is home to all the species mentioned above.

Geology

The Iberian Peninsula contains rocks from every age from Ediacaran to Recent, and almost every kind of rock is represented. World class mineral deposits can also be found there. The core of the Iberian Peninsula consists of a Hercynian cratonic block known as the Iberian Massif.

On the northeast this is bounded by The Pyrenean fold belt, and on the southeast it is bounded by the Betic Foldchain. These twofold chains are part of the Alpine belt. To the west, the peninsula is delimited by the continental boundary formed by the magma poor opening of the Atlantic Ocean. The Hercynian Foldbelt is mostly buried by Mesozoic and Tertiary cover rocks to the east, but nevertheless outcrops through the Iberian Chain and the Catalonian Coastal Ranges.

Prehistory

The Iberian Peninsula has been inhabited for at least 1,000,000 years as remains found in the sites at Atapuerca demonstrate. Among these sites is the cave of Gran Dolina, where six hominin skeletons, dated between 780,000 and one million years ago, were found in 1994. Experts have debated whether these skeletons belong to the species *Homo erectus*, *Homo heidelbergensis*, or a new species called *Homo antecessor*.

Around 200,000 BC, during the Lower Paleolithic period, Neanderthals first entered the Iberian Peninsula. Around 70,000 BC, during the Middle Paleolithic period, the last ice age began and the Neanderthal Mousterian culture was established. Around 35,000 BC, during the Upper Paleolithic, the Neanderthal Châtelperronian cultural period began. Emanating from Southern France this culture extended into Northern Iberia. It continued to exist until around 28,000 BC when Neanderthal man faced extinction. At about the 40th millennium BC Modern Humans entered the Iberian peninsula, coming from Southern France.

Here, this genetically homogeneous population (characterized by the M173 mutation in the Y chromosome), developed the M343 mutation, giving rise to the R1b Haplogroup, still the most common in modern Portuguese and Spanish males. In Iberia, Modern Humans developed a series of different cultures, such as the Aurignacian, Gravettian, Solutrean and Magdalenian cultures, some of them characterized by complex forms of Paleolithic art.

Neolithic

During the Neolithic expansion, various megalithic cultures developed in Iberia. An open seas navigation culture from the east Mediterranean, called the Cardium culture, also extended its influence to the eastern coasts of Iberia, possibly as early as the 5th millennium BC These people may have had some relation to the subsequent development of the Iberian civilization.

Chalcolithic

In the Chalcolithic or Copper Age (c. 3000 BC in Iberia) a series of complex cultures developed, which would give rise to the first civilizations in Iberia and to extensive exchange networks reaching to the Baltic, the Middle East and North Africa. At about 2150 BC the Bell Beaker culture intruded into Chalcolithic Iberia, being of Central European origin.

Bronze Age

Bronze Age cultures developed beginning c.1800 BC, when the civilization of Los Millares was followed by that of El Argar. From this centre, bronze technology spread to other areas, such as those of the Bronze of Levante, South-Western Iberian Bronze and Cogotas I.

In the Late Bronze Age the urban civilisation of Tartessos developed in the area of modern western Andalusia, characterized by Phonation influence and using the Tartessian script for its Tartessian language, a language isolate not related to the Iberian language.

Early in the first millennium BC, several waves of Pre-Celts and Celts migrated from central Europe, thus partially changing the ethnic landscape of Iberia into Indo-European space in its northern and western regions.

Proto-history

By the Iron Age, starting in the 7th century BC, the global panorama in Iberia was one of complex agrarian and urban civilizations, either Pre-Celtic or Celtic (such as the Lusitanians, the Celtiberians, the Gallaeci, the Astur, or the Celtici, amongst others), the cultures of the Iberians in the eastern and southern zones of Iberia and the cultures of the Aquitanian in the western portion of the Pyrenees.

The seafaring Phoenicians, Greeks and Carthaginians successively settled along the Mediterranean coast and founded trading colonies there over a period of several centuries. Around 1100 BC Phonation merchants founded the trading colony of Gadir or Gades (modern day Cadiz) near Tartessos. In the 8th century BC the first Greek colonies, such as Emporion (modern Empuries), were founded along the Mediterranean coast on the East, leaving the south coast to the Phoenicians. The Greeks are responsible for the name Iberia, after the river Iber (Ebro). In the 6th century BC the Carthaginians arrived in Iberia while struggling with the Greeks for control of the Western Mediterranean. Their most important colony was Carthago Nova (Latin name of modern day Cartagena).

History

Roman Iberia

In 219 BC, the first Roman troops invaded the Iberian Peninsula, during the Second Punic war against the Carthaginians, and annexed it under Augustus after two centuries of war with the Celtic and Iberian tribes and the Phonation, Greek and Carthaginian colonies, resulting in the creation of the province of Hispania. It was divided into Hispania

Ulterior and Hispania Citerior during the late Roman Republic, and during the Roman Empire, it was divided into Hispania Taraconensis in the northeast, Hispania Baetica in the south and Lusitania in the southwest.

Hispania supplied the Roman Empire with food, olive oil, wine and metal. The emperors Trajan, Hadrian and Theodosius I, the philosopher Seneca and the poets Martial and Lucan were born from families living in Iberia.

Germanic Iberia

In the early 5th century, Germanic tribes invaded the peninsula, namely the Suevi, the Vandals (Silingi and Hasdingi) and their allies, the Sarmatian Alans. Only the kingdom of the Suevi (Quadi and Marcomanni) would endure after the arrival of another wave of Germanic invaders, the Visigoths, who conquered all of the Iberian peninsula and expelled or partially integrated the Vandals and the Alans. The Visigoths eventually conquered the Suevi kingdom and its capital city Bracara (modern day Braga) in 584-585. They would also conquer the province of the Byzantine Empire (552-624) of Spania in the south of the peninsula and the Balearic Islands.

Islamic Iberia

In 711 AD, a North African Moorish Umayyad army invaded Visigothic Christian Hispania. Under their leader Tariq ibn-Ziyad, they landed at Gibraltar and brought most of the Iberian Peninsula under Islamic rule in an eight-year campaign. Al-Andalus (Arabic : Land of the Vandals) is the Arabic name given the Iberian Peninsula by its Muslim conquerors and its subsesquent inhabitants.

From the 8th to the 15th centuries, parts of the Iberian peninsula were ruled by the Moors (mainly Berber and Arab) who had crossed over from North Africa.

Reconquest

Many of the ousted Gothic nobles took refuge in the unconquered north Asturian highlands. From there they aimed to reconquer their lands from the Moors: this war of reconquest is known as the Reconquista. Christian and Muslim kingdoms fought and allied among themselves. The Muslim taifa kings competed in patronage of the arts, the Way of Saint James attracted pilgrims from all Western Europe and the Jewish population of Iberia set the basis of Sephardic culture.

In medieval times the peninsula housed many small states including Castile, Aragon, Navarre, Leon and Portugal. The peninsula was part of the Islamic Almohad empire until they were finally uprooted. The last major Muslim stronghold was Granada which was eliminated by a combined Castilian and Aragonese force in 1492.

Post Reconquest

The small states gradually amalgamated over time, with the exception of Portugal, even if for a brief period (1580–1640) the whole peninsula was united politically under the Iberian Union. After that point the modern position was reached and the peninsula now consists of the countries of Spain and Portugal (excluding their islands-the Portuguese Azores and Madeira Islands and the Spanish Canary Islands and Balearic Islands; and the Spanish exclaves of Ceuta and Melilla), Andorra, French Cerdagne and Gibraltar.

4

Union of Soviet Socialist Republics: Now a World Power

The Union of Soviet Socialist Republics, also known as the Soviet Union, was a constitutionally socialist state that existed on the territory of most of the former Russian Empire in Eurasia between 1922 and 1991.

The Soviet Union had a single-party political system dominated by the Communist Party. Nominally a union of Soviet republics, of which there were 15 after 1956, with the capital in Moscow, *de facto* the Soviet Union was a highly centralized state with a planned economy. The security agency KGB actively oversaw much of Soviet society.

The union was founded in December 1922, when the Russian SFSR, which formed during the Russian Revolution of 1917 and emerged victorious in the ensuing Russian Civil War, unified with the Transcaucasian, Ukrainian and Belorussian SSRs. After the death of Vladimir Lenin, the first Soviet leader, power was eventually consolidated by Joseph Stalin, who led the country through a large-scale industrialization with command economy and political repression. During World War II, in June 1941, the Soviet Union was attacked by Germany, a country it had signed a non-aggression pact with. After four years of warfare, the Soviet Union emerged as one of the world's two superpowers, extending its influence into much of Eastern Europe and beyond. The Cold War, a global ideological and political struggle between the Soviet Union and its satellites from the Eastern Bloc on the one side and the United States and its allies on the other side, which the Soviet bloc, hit by economic standstill, ultimately lost, marked the post-war period. In the late 1980s the last Soviet leader Mikhail Gorbachev tried to reform the state with his policies of perestroika and glasnost, but the Soviet Union ultimately collapsed and was formally dissolved in December 1991 after the abortive August coup attempt. Since then the Russian Federation has been exercising its rights and fulfilling its obligations.

Geography

The Soviet Union, with 22,402,200 square kilometres (8,649,500 sq mi), was the world's largest state. Covering a sixth of the world's inhabited land, its size was comparable to that of North America. The western part (in Europe) accounted for a quarter of the country's area, and was the country's cultural and economic center. The eastern part (in Asia) extended to the Pacific Ocean to the east and Afghanistan to the south, and was much less populated than the European part. It was over 10,000 kilometres (6,200 mi) across (11 time zones) and almost 5,000 kilometres (3,100 mi) north to south. Its five climatic zones were tundra, taiga, steppes, desert, and mountains.

The Soviet Union had the world's longest border, measuring over 60,000 kilometres (37,000 mi). Two thirds of the Soviet border was coastline of the Arctic Ocean. Across the Bering Strait was the United States. The Soviet Union bordered Afghanistan, China, Czechoslovakia, Finland, Hungary, Iran, Mongolia, North Korea, Norway, Poland, Romania, and Turkey at the end of WWII.

The Soviet Union's longest river was the Irtysh. The Soviet Union's highest mountain was Communism Peak (today's Ismail Samani Peak) in Tajikistan at 7,495 metres (24,590 ft). The world's largest lake, the Caspian Sea, lay mainly in the Soviet Union. The world's deepest lake, Lake Baikal, was in the Soviet Union.

Demographics

The Soviet Union was one of the world's most ethnically diverse countries, with more than 100 distinct ethnic groups within its borders. The total population was estimated at 293 million in 1991. It was the third most populous nation (after China and India) for decades. There were 23 cities with more than one million people each in the Soviet Union in 1989. The country's largest city and capital was Moscow with nine million inhabitants.

Ethnic Groups

As a 1990 estimate, the majority of the population were Russians (50.78%), followed by Ukrainians (15.45%) and Uzbeks (5.84%).

Some nationality groups came into the empire voluntarily, others were brought in by force. Russians, Belarusians, and Ukrainians shared close cultural ties while, other subjects of the empire did not. Due to multiple nationalities located in the same territory, national antagonisms developed over the years.

For many years, Soviet leaders maintained that the underlying causes of conflict between nationalities had been eliminated and that the Soviet Union consisted of a family of nations living harmoniously together. In the 1920s and early 1930s, the government conducted a policy of korenizatsiya (indigenization) of local governments in an effort to recruit non-Russians into the new Soviet political institutions and to reduce the conflict between Russians and the minority nationalities.

To increase literacy and mass education, the Soviets encouraged development and publication in many of the languages of minority groups. Russian became a required *subject* of study in all Soviet schools in 1938; however, in mainly non-Russian areas the chief language of instruction was the local language. This led to widespread bilingualism in the educated population, though smaller nationalities were often linguistically assimilated, in which the members of the nationality lost their historic language.

Religion

The Soviet Union was officially secular, supported atheism in schools, and suppressed religion. The state was separated from church by the Decree of Council of People's Commissars on January 23, 1918. Two-thirds of the Soviet population lacked religious belief while one-third of the people professed religious belief. Christianity and Islam had the most believers. About half of the people, including members of the CPSU and high-level government officials, professed atheism. Government persecution of Christianity continued undiminished until the fall of the Communist government. Only 500 churches, out of the 54,000 before the revolution, remained open in 1941. The role of religion in the daily lives of Soviet citizens varied greatly and was far less integral in city dwellers where Party control was optimum.

Language

While everyone could use their own language, Russian was the official and dominant language in the Soviet Union. It was used in industry, military, party, and state management.

Life Expectancy

After the communist takeover of power the life expectancy for all age groups went up. The trend continued into the 60s, when the life expectancy in the Soviet Union went beyond the life expectancy in the United States. In 1964 the trend reversed. Life expectancy went down dramatically for men because of alcohol abuse and poor health care.

Emigration

Despite the strict rules that would prevent emigration, there were many Soviet citizens who wanted to seek their fortunes elsewhere. The desire to emigrate elsewhere applied especially to the Jewish community of USSR, as they often felt alienated by the anti-Semitic attitudes in society and sometimes even anti-Jewish campaigns of the Soviet State.

History

The last Russian Tsar, Nicholas II, ruled until March 1917, when the Russian Empire was overthrown and a short-lived Russian provisional government took power, to be overthrown in November 1917 by Vladimir Lenin. From 1917 to 1922, the predecessor to the Soviet Union was the Russian Soviet Federative Socialist Republic (RSFSR), which was an independent country, as were other Soviet republics at the time. The Soviet Union was officially established in December 1922 as the union of the Russian (colloquially known as Bolshevist Russia), Ukrainian, Belarusian, and Transcaucasian Soviet republics ruled by Bolshevik parties.

Revolution and the Foundation of a Soviet State

Modern revolutionary activity in the Russian Empire began with the Decembrist Revolt of 1825, and although serfdom was abolished in 1861, its abolition was achieved on terms unfavorable to the peasants and served to encourage revolutionaries. A parliament—the State Duma—was established in 1906 after the Russian Revolution of 1905, but the Tsar resisted attempts to move from absolute to constitutional monarchy. Social unrest continued and was aggravated during World War I by military defeat and food shortages in major cities.

A spontaneous popular uprising in Saint Petersburg, in response to the wartime decay of Russia's economy and morale, culminated in the "February Revolution" and the toppling of the imperial government in March 1917. The tsarist autocracy was replaced by the Provisional Government, whose leaders intended to conduct elections to Russian Constituent Assembly and to continue participating on the side of the Entente in World War I.

At the same time, to ensure the rights of the working class, workers' councils, known as Soviets, sprang up across the country. The Bolsheviks, led by Vladimir Lenin, pushed for socialist revolution in the Soviets and on the streets. In November 1917, during the "October Revolution," they seized power from the Provisional Government. In December, the Bolsheviks signed an armistice with the Central Powers. But, by February 1918, fighting had resumed. In March, the Soviets quit the war for good and signed the Treaty of Brest-Litovsk.

Only after the long and bloody Russian Civil War was the new Soviet power secure. The civil war between the Reds and the Whites started in 1917 and ended in 1923. It included foreign intervention, the execution of Nicholas II and his family and the famine of 1921, which killed about 5 million. In March

1921, during a related conflict with Poland, the Peace of Riga was signed and split disputed territories in Belarus and Ukraine between the Republic of Poland and Soviet Russia. The Soviet Union had to resolve similar conflicts with the newly established Republic of Finland, the Republic of Estonia, the Republic of Latvia, and the Republic of Lithuania.

Unification of the Soviet Republics

On December 28, 1922, a conference of plenipotentiary delegations from the Russian SFSR, the Transcaucasian SFSR, the Ukrainian SSR and the Byelorussian SSR approved the Treaty of Creation of the USSR and the Declaration of the Creation of the USSR, forming the Union of Soviet Socialist Republics. These two documents were confirmed by the 1st Congress of Soviets of the USSR and signed by heads of delegations – Mikhail Kalinin, Mikha Tskhakaya, Mikhail Frunze and Grigory Petrovsky, Aleksandr Chervyakov respectively on December 30, 1922. On February 1, 1924, the USSR was recognized by the British Empire. Also in 1924, a Soviet Constitution was approved, legitimizing the December 1922 union of the Russian SFSR, the Ukrainian SSR, the Belarusian SSR, and the Transcaucasian SFSR to form the "Union of Soviet Socialist Republics" (USSR).

The intensive restructuring of the economy, industry and politics of the country began in the early days of Soviet power in 1917. A large part of this was performed according to Bolshevik Initial Decrees, documents of the Soviet government, signed by Vladimir Lenin. One of the most prominent breakthroughs was the GOELRO plan, that envisioned a major restructuring of the Soviet economy based on total electrification of the country. The Plan was developed in 1920 and covered a 10-to 15-year period. It included construction of a network of 30 regional power plants, including ten large hydroelectric power plants, and numerous electric-powered large industrial enterprises. The Plan became the prototype for subsequent Five-Year Plans and was basically fulfilled by 1931.

Stalin's Rule

From its beginning years, government in the Soviet Union was based on the one-party rule of the Communist Party

(Bolsheviks). After the economic policy of War Communism during the Civil War, the Soviet government permitted some private enterprise to coexist with nationalized industry in the 1920s and total food requisition in the countryside was replaced by a food tax.

Soviet leaders argued that one-party rule was necessary because it ensured that 'capitalist exploitation' would not return to the Soviet Union and that the principles of Democratic Centralism would represent the people's will. Debate over the future of the economy provided the background for Soviet leaders to contend for power in the years after Lenin's death in 1924. Initially, Lenin was to be replaced by a "troika" composed of Grigory Zinoviev of Ukraine, Lev Kamenev of Moscow, and Joseph Stalin of Georgia.

On 3 April 1922, Stalin had been named the General Secretary of the Communist Party of the Soviet Union. Lenin had appointed Stalin to be the head of the Workers' and Peasants' Inspectorate, known by the acronym Rabkrin, which gave Stalin considerable power. By gradually consolidating his influence and isolating and out-maneuvering his rivals within the party, Stalin became the undisputed leader of the Soviet Union by the end of the 1920s establishing totalitarian rule. In October 1927, Grigory Zinoviev and Leon Trotsky were expelled from the Central Committee and forced into exile.

In 1928, Stalin introduced the First Five-Year Plan for building a socialist economy. While encompassing the internationalism expressed by Lenin throughout the course of the Revolution, it also aimed for building socialism in one country. In industry, the state assumed control over all existing enterprises and undertook an intensive program of industrialization; in agriculture collective farms were established all over the country.

Famines occurred, causing millions of deaths and surviving kulaks were politically persecuted and many sent to Gulags to do forced labour. Social upheaval continued in the mid-1930s. Stalin's Great Purge resulted in execution or detainment of many "Old Bolsheviks" who had participated in the October Revolution with Lenin. According to the declassified Soviet archives, during the Great Purge in 1937 and 1938, the NKVD arrested more than one and a half million people, of whom 681,692 were shot – an average of 1,000 executions a day. Yet despite the turmoil of the mid-to late 1930s, the Soviet Union developed a powerful industrial economy in the years before World War II.

The 1930s

The early 1930s saw closer cooperation between the West and the USSR. From 1932 to 1934, the Soviet Union participated in the World Disarmament Conference. In 1933, diplomatic relations between the United States and the USSR were established. In September 1934, the Soviet Union joined the League of Nations. After the Spanish Civil War broke out in 1936, the USSR actively supported the Republican forces against the Nationalists. The Nationalists were supported by Fascist Italy and Nazi Germany.

In December 1936, Stalin unveiled a new Soviet Constitution. This constitution provided economic rights not included in constitutions in the western democracies. The constitution was seen as a personal triumph for Stalin, who on this occasion was described by Pravda as "genius of the new world, the wisest man of the epoch, the great leader of communism." By contrast, western historians and historians from former Soviet occupied countries have seen the constitution as a meaningless propaganda document.

The late 1930s saw a shift towards the Axis powers. In 1938 and 1939, armed forces of the USSR won several decisive victories during border clashes with the armed forces of the Japanese Empire. In 1938, after the United Kingdom and France concluded the Munich Agreement with Germany, the USSR dealt with Germany as well.

World War II

The USSR dealt with Germany both militarily and economically during extensive talks and by concluding the German-Soviet Nonaggression Pact and the German–Soviet Commercial Agreement. The conclusion of the nonaggression pact made possible the Soviet occupation of Lithuania, Latvia,

Estonia, Bessarabia, northern Bukovina, and eastern Poland. In late November of the same year, unable to force the Republic of Finland into agreement to move its border 25 kilometres (16 mi) back from Leningrad by diplomatic means, Stalin ordered the invasion of Finland. On April 1941, USSR signed the Soviet–Japanese Neutrality Pact with the Empire of Japan, recognizing the territorial integrity of Manchukuo, a Japanese puppet state.

Although it has been debated whether the Soviet Union had the intention of invading Germany once it was strong enough, Germany itself broke the treaty and invaded the Soviet Union on 22 June 1941 and started what was known in the USSR as the "Great Patriotic War." The Red Army stopped the initial German offensive during the Battle of Moscow. The Battle of Stalingrad, which lasted from late 1942 to early 1943, was a major defeat for the Germans and became a major turning point of the war. After Stalingrad, Soviet forces drove through Eastern Europe to Berlin before Germany surrendered in 1945. The same year, the USSR, in fulfilment of its agreement with the Allies at the Yalta Conference, denounced the Soviet–Japanese Neutrality Pact in April 1945 and invaded Manchukuo and other Japan-controlled territories on August 9, 1945. This conflict ended with a decisive Soviet victory, contributing to the unconditional surrender of Japan and the end of World war II. The Soviet Union lost around 27 million people in the war. Although ravaged by the war, the Soviet Union emerged victorious from the conflict and became an acknowledged military superpower.

Once denied diplomatic recognition by the free world, the Soviet Union had official relations with practically all nations of the world by the late 1940s. The Soviet Union also had progressed from being an outsider in international organizations and negotiations to being one of the arbiters of the world's fate after World War II. A member of the United Nations at its foundation in 1945, the Soviet Union became one of the five permanent members of the UN Security Council which gave it the right to veto any of its resolutions.

The Soviet Union emerged from World War II as one of the world's two superpowers, a position maintained for four decades through its hegemony in Eastern Europe, military strength, economic strength, aid to developing countries, and scientific research, especially into space technology and weaponry. The Soviet Union's growing influence abroad in the postwar years helped lead to a Communist system of states in Eastern Europe united by military and economic agreements.

The Cold War

During the immediate postwar period, the Soviet Union first rebuilt and then expanded its economy, while maintaining its strictly centralized control. The Soviet Union aided post-war reconstruction in the countries of Eastern Europe while turning them into Soviet satellite states, founded the Warsaw Pact in 1955. The Council for Mutual Economic Assistance (Comecon), 1949–1991, was an economic organization of communist states and a kind of Eastern Bloc equivalent to—but more geographically inclusive than—the European Economic Community. Later, the Comecon supplied aid to the eventually victorious Communists in the People's Republic of China, and saw its influence grow elsewhere in the world. Meanwhile, the rising tension of the Cold War turned the Soviet Union's wartime allies, the United Kingdom and the United States, into enemies.

Post-Stalin Period

Stalin died on March 5, 1953. In the absence of an acceptable successor, the highest Communist Party officials opted to rule the Soviet Union jointly. Nikita Khrushchev, who had won the power struggle by the mid-1950s, denounced Stalin's use of repression in 1956 and eased repressive controls over party and society. This was known as de-Stalinization.

Moscow considered Eastern Europe to be a buffer zone for the forward defense of its western borders and ensured its control of the region by transforming the East European countries into satellite states. Soviet military force was used to suppress anti-communist uprisings in Hungary and Poland in 1956. In the late 1950s, a confrontation with China regarding the USSR's rapprochement with the West and what Mao perceived as Khrushchev's revisionism led to the Sino-Soviet split. This resulted in a break throughout the global Communist movement and Communist regimes in Albania and Cambodia choosing to ally with China in place of the USSR. During this period, the Soviet Union continued to realize scientific and technological pioneering exploits; to launch the first artificial satellite, Sputnik 1; a living dog, Laika; and later, the first human being, Yuri Gagarin, into Earth's orbit. Valentina Tereshkova was the first woman in space aboard Vostok 6 on June 16, 1963, and Alexey Leonov became the first person to walk in space on March 18, 1965. Khrushchev's reforms in agriculture and administration, however, were generally unproductive. During the same period, a tense confrontation between the Soviet Union and the United States over the Soviet deployment of nuclear missiles in Cuba sparked the Cuban Missile Crisis in 1962. Khrushchev was retired from power in 1964.

Following the ousting of Khrushchev, another period of rule by collective leadership ensued, lasting until Leonid Brezhnev established himself in the early 1970s as the preeminent figure in Soviet political life. In 1968 the Soviet Union and members of its Warsaw Pact allies invaded Czechoslovakia to halt the Prague Spring reforms.

Brezhnev presided over a period of *Detente* with the West while at the same time building up Soviet military strength.

In October 1977, the third Soviet Constitution was unanimously adopted. The prevailing mood of the Soviet leadership at the time of Brezhnev's death in 1982 was one of aversion to change. The long period of Brezhnev's rule had come to be dubbed one of "standstill", with an aging and ossified top political leadership.

Reforms of Gorbachev and Collapse of the Soviet Union

Two developments dominated the decade that followed: the increasingly apparent crumbling of the Soviet Union's economic and political structures, and the patchwork attempts at reforms to reverse that process. Kenneth S. Deffeyes argued in *Beyond Oil* that the Reagan administration encouraged Saudi Arabia to lower the price of oil to the point where the Soviets could not make a profit from selling their oil, so that the USSR's hard currency reserves became depleted. After the rapid succession of Yuri Andropov and Konstantin Chernenko, transitional figures with deep roots in Brezhnevite tradition, beginning in

1985 Mikhail Gorbachev made significant changes in the economy and the party leadership. His policy of *glasnost* freed public access to information after decades of heavy government censorship. With the Soviet Union in bad economic shape and its satellite states in eastern Europe abandoning communism, Gorbachev moved to end the Cold War.

After Mikhail Gorbachev succeeded Konstantin Chernenko as General Secretary of the CPSU in 1985, he introduced many changes in Soviet foreign policy and in the economy of the USSR.

In 1988, the Soviet Union abandoned its nine-year war with Afghanistan and began to withdraw forces from the country. In the late 1980s, Gorbachev refused to send military support to defend the Soviet Union's former satellite states, resulting in multiple communist regimes in those states being forced from power. With the tearing down of the Berlin Wall and with East Germany and West Germany pursuing unification, the Iron Curtain took the final blow.

In the late 1980s, the constituent republics of the Soviet Union started legal moves towards or even declaration of sovereignty over their territories, citing Article 72 of the USSR Constitution, which stated that any constituent republic was free to secede. On April 7, 1990, a law was passed allowing a republic to secede if more than two-thirds of that republic's residents vote for secession on a referendum. Many held their first free elections in the Soviet era for their own national legislatures in 1990. Many of these legislatures proceeded to produce legislation contradicting the Union laws in what was known as the "War of Laws."

In 1989, the Russian SFSR, which was then the largest constituent republic (with about half of the population) convened a newly elected Congress of People's Deputies. Boris Yeltsin was elected the chairman of the Congress. On June 12, 1990, the Congress declared Russia's sovereignty over its territory and proceeded to pass laws that attempted to supersede some of the USSR's laws. The period of legal uncertainty continued throughout 1991 as constituent republics slowly became de facto independent.

A referendum for the preservation of the USSR was held on March 17, 1991, with the majority of the population voting for preservation of the Union in nine out of 15 republics. The referendum gave Gorbachev a minor boost, and, in the summer of 1991, the New Union Treaty was designed and agreed upon by eight republics which would have turned the Soviet Union into a much looser federation.

The signing of the treaty, however, was interrupted by the August Coup—an attempted coup d'état against Gorbachev by hardline Communist Party members of the government and the KGB, who sought to reverse Gorbachev's reforms and reassert the central government's control over the republics. After the coup collapsed, Yeltsin—who had publicly opposed it—came out as a hero while Gorbachev's power was effectively ended. The balance of power tipped significantly towards the republics. In August 1991, Latvia and Estonia immediately declared restoration of full independence (following Lithuania's 1990 example), while the other twelve republics continued discussing new, increasingly looser, models of the Union.

On December 8, 1991, the presidents of Russia, Ukraine and Belarus signed the Belavezha Accords which declared the Soviet Union dissolved and established the Commonwealth of Independent States (CIS) in its place. While doubts remained over the authority of the Belavezha Accords to dissolve the Union, on December 21, 1991, the representatives of all Soviet republics except Georgia, including those republics that had signed the Belavezha Accords, signed the Alma-Ata Protocol, which confirmed the dismemberment and consequential extinction of the USSR and restated the establishment of the CIS. The summit of Alma-Ata also agreed on several other practical measures consequential to the extinction of the Union. On December 25, 1991, Gorbachev yielded to the inevitable and resigned as the president of the USSR, declaring the office extinct. He turned the powers that until then were vested in the presidency over to Boris Yeltsin, president of Russia.

The following day, the Supreme Soviet, the highest governmental body of the Soviet Union, recognized the bankruptcy and collapse of the Soviet Union and dissolved itself. This is generally recognized as the official, final dissolution of the Soviet Union as a functioning state. Many organizations such as the Soviet Army and police forces continued to remain in place in the early months of 1992 but were slowly phased out and either withdrawn from or absorbed by the newly independent states.

Following the dissolution of the Soviet Union on December 25, 1991, Russia was internationally recognized to be the legal successor to the Soviet state on the international stage. To that end, Russia voluntarily accepted all Soviet foreign debt, and claimed overseas Soviet properties as its own. Since then the Russian Federation has been exercising its rights and fulfilling its obligations.

Government and Politics

There were three power hierarchies in the Soviet Union: the state systems of Soviets and ministries as well as the Communist Party of the Soviet Union (CPSU), the only legal party and the ultimate policymaker in the country, known under the name of the All-Union Communist Party (Bolsheviks) before the 19th Party Congress in 1952.

At the top of the Communist Party was the Central Committee, elected at Party Congresses and Conferences. The Central Committee in turn voted for a Politburo (called Presidium between 1952-1966), Secretariat and Secretary General (First Secretary in 1953-1966), position that came to prominence with Stalin's rise to power. Depending on the degree of power consolidation, it was either the Politburo as a collective body or the Secretary General, who always was one of the Politburo members, that effectively led the party and the country. They weren't controlled by the mass of the party membership, as the key principle of the party organization was democratic centralism, demanding strict subordination to the higher bodies, and the elections went uncontested, endorsing the candidates proposed from above.

The Supreme Soviet (known before 1936 as the Central Executive Committee), nearly unanimously voted for by the population in uncontested and less than secret elections, nominally the highest state body for most of the Soviet history, *de facto* was a rubber stamp institution, approving and implementing all decisions imposed on it by the party. It elected a Presidium to wield its power between plenary sessions, ordinarily held twice a year, and appointed the Supreme Court, the Procurator General and the Council of Ministers (known before 1946 as the Council of People's Commissars), headed by the Chairman (Premier) and managing an enormous bureaucracy responsible for the administration of the economy and society.

State and party structures of the constituent republics largely emulated the structure of the central institutions, although the Russian SFSR, unlike the other constituent republics, for most of its history had no republican branch of the CPSU, being ruled directly by the union-wide party until 1990. Local authorities were organized likewise into party committees, local Soviets and executive committees. While the state system was nominally federal, the party was unitary.

The Communist Party maintained its dominance over the state largely through its control over the system of appointments. All senior government officials and most deputies of the Supreme Soviet were members of the CPSU, the more important they were the higher their position in the party hierarchy. Of the party heads themselves, Stalin in 1941-1953 and Khrushchev in 1958-1964 were Premiers. Upon the forced retirement of Khrushchev the party head became prohibited from this kind of double membership, but the later Secretaries General for at least some part of their tenure in office occupied the position of the Chairman of the Presidium of the Supreme Soviet, nominal head of state, albeit largely ceremonial. The institutions at lower levels were overseen and at times supplanted by primary party organizations. In practice, however, the control the party was able to exercise over the state bureaucracy, particularly after the death of Stalin, was far from total, with the state bureaucracy pursuing different interests, at times in conflict with the party. Neither was the party itself monolithic from top to bottom, although factions were officially banned.

The state security police (the KGB and its predecessor agencies) played important role in the Soviet politics. It was instrumental in the Stalinist terror. After the death of Stalin the state security police was brought under strict party control. Under Yury Andropov, KGB chairman in 1967-1982 and Secretary General in 1982-1983, the KGB, engaging in the suppression of political dissent and maintaining an extensive network of informers, reasserted itself as a political actor to some extent independent of the party-state structure, culminating in the anti-corruption campaign targeting high party officials in the late 1970s-early 1980s.

The Soviet constitutions, which were promulgated in 1918, 1924, 1936 and 1977, didn't limit state power. No formal separation of powers existed between the Party, Supreme Soviet and Council of Ministers, the fusion of executive and legislative functions was pervasive. The system was governed less by statute than by informal conventions. No settled mechanism of leadership succession existed. Bitter and at times deadly power strugle took place in the Politburo after the deaths of Lenin and Stalin, as well as after Khrushchev's dismissal, itself due to a coup in the Central Committee. With the only exception of Khrushchev, all Soviet party leaders before Gorbachev died in office.

In 1988-1990, facing considerable opposition, Secretary General Mikhail Gorbachev enacted reforms shifting power away from the highest bodies of the party and making the Supreme Soviet less dependent on them. The Congress of People's Deputies was established, majority of whose members were directly elected by the population in competitive elections held in March 1989. The Congress now elected the Supreme Soviet, which became a full-time parliament, much stronger than before, and, although still being largely conservative, for the first time since the 1920s refused to rubber-stamp proposals from the party and Council of Ministers. In 1990 Gorbachev introduced and assumed the position of the President of the Soviet Union, concentrated power in his executive office, independent of the party, and subordinated the government, now renamed Cabinet of Ministers, to himself. Tensions were growing between the union-wide authorities under Gorbachev, reformists, led in Russia by Boris Yeltsin and controlling the newly elected Supreme Soviet of the Russian SFSR, and Communist Party hardliners. On August 19-21, 1991, a group of hardliners staged an abortive coup attempt. Following its failure Gorbachev resigned as Secretary General, only remaining Soviet President for the final months of the existence of the union, and the USSR Supreme Soviet suspended the CPSU.

Judicial System

The judiciary was not independent from the other branches of government. The Supreme Court supervised the lower courts and applied the law as established by the Constitution or as interpreted by the Supreme Soviet. The Constitutional Oversight Committee reviewed the constitutionality of laws and acts. The Soviet Union utilized the inquisitorial system of Roman law, where judge, procurator, and defense attorney work collaboratively to establish the truth.

Soviet Socialist Republics

Constitutionally, the Soviet Union was a federation of Soviet Socialist Republics (SSRs) and the Russian Soviet Federative Socialist Republic (RSFSR), although the rule of the highly cenralized Communist Party made Soviet federalism merely nominal. The Treaty on the Creation of the USSR was signed in December 1922 by four founding republics, the RSFSR, Transcaucasian SFSR, Ukrainian SSR and Belorussian SSR. In 1924, during the national delimitation in Central Asia, the Uzbek and Turkmen SSRs were formed from parts of the RSFSR's Turkestan ASSR and two Soviet dependencies, the Khorezm and Bukharan SSR. In 1929 the Tajik SSR was split off from the Uzbek SSR.

With the constitution of 1936 the constituents of the Transcaucasian SFSR, namely the Georgian, Armenian and Azerbaijan SSRs, were elevated to union republics, whereas the Kazakh and Kirghiz SSRs were split off from the RSFSR. In August 1940 the Soviet Union formed the Moldavian SSR from parts of the Ukrainian SSR and parts of Bessarabia annexed from Romania, as well as annexed the Baltic states as the Estonian, Latvian and Lithuanian SSRs. The Karelo-Finnish SSR was split off from the RSFSR in March 1940 and merged back in 1956. Between July 1956 and September 1991 there were 15 union republics.

In March 1990 the newly elected Supreme Soviet of the Lithuanian SSR declared independence, followed by the Georgian Supreme Soviet in April 1991. Although the symbolic right of the union republics to secede was nominally guaranteed by the constitution and the union treaty, the union authorities at first refused to recognize it. After the August coup attempt most of the other republics followed suit. The Soviet Union ultimately recognized the secession of Estonia, Latvia and Lithuania on September 6, 1991. The remaining republics were recognized as independent with the union's final dissolution in December 1991.

Economy

The Soviet Union became the first country that adopted a planned economy, whereby production and distrubution of goods were to be centralized and directed by the government. The first Bolshevik experience with command economy was the policy of War Communism, involving nationalization of industry, centralized distribution of output, coercive requisition of agricultural production, and attempts to eliminate money circulation, as well as private enterprises and free trade. As it had aggravated a severe economic collapse caused by the war, in 1921 Lenin replaced War Communism with the New Economic Policy (NEP), legalizing free trade and private ownership of smaller businesses. The economy subsequently recovered fairly quickly.

Following a lengthy debate among the members of Politburo over the course of economic development, by 1928-1929, upon gaining the upper hand in the power struggle, Joseph Stalin had abandoned the NEP and pushed for full central planning, starting forced collectivization of agriculture and enacting draconian labour legislation. The resources were mobilized for rapid industrialization, which greatly expanded Soviet capacity in heavy industry and capital goods during the 1930s. Preparation for war was one of the main driving forces behind industrialization, mostly due to distrust of the outside capitalistic world. As a result, the USSR was transformed from a largely agrarian economy into a great industrial power, and the basis was provided for its emergence as a superpower after recovering from the World War II. During the war the Soviet economy and infrastructure suffered massive devastation and subsequently required extensive reconstruction.

A wide range of industries constituted the Soviet industrial sector in the later decades, including machine-building and metal-working, metallurgy, chemicals, petroleum and natural gas, coal mining, forestry, defense industry, textiles, food processing, and construction. By 1980, the Soviet Union had the world's second largest industrial capacity with 20 percent of total world industrial output, leading the world in producing oil, cast iron, steel, coke, mineral fertilizers, locomotives, tractors, and cement.

By the early 1940s, the Soviet economy had become relatively autarkic; for most of the period up until the creation of Comecon, only a very small share of domestic products were traded internationally. After the creation of the Eastern Bloc, external trade rose rapidly. Still the influence of the world economy on the USSR was limited by fixed domestic prices and state monopoly on the foreign trade. Grain and sophisticated consumer manufactures became major import articles from around 1960s. Petroleum and petroleum products, natural gas, metals, wood, agricultural products, and a variety of manufactured goods, primarily machinery, arms and military equipment, were exported from the country. In the 1970s-1980s, the Soviet Union heavily relied on fossil fuel exports to earn hard currency. At the peak level in 1988, it was the largest producer and second largest exporter of crude oil, surpassed only by Saudi Arabia.

The Soviet Union placed great emphasis on science and technology within its economy, however the most remarkable Soviet successes in technology, such as producing the world's first space satellite, typically were the military responsibility. During the arms race of the Cold War the Soviet economy was burdened by military expenditures, heavily lobbied by the powerful bureaucracy dependent on the arms industry. At the same time the Soviet Union became the largest arms exporter to the Third World. Significant amounts of the Soviet resources during the Cold War were allocated in aid to the other communist countries.

Since the 1930s and until its collapse in the late 1980s, the way the Soviet economy operated had remained essentially unchanged. The economy was formally directed by central planning, carried out by Gosplan and organized into five-year plans. In practice, however, the plans were highly aggregated and provisional, subject to *ad hoc* intervention by superiors. All key economic decisions were taken by the political leadership. Allocated resources and plan targets were normally denominated in rubles rather than in physical goods. Credits were discouraged, but widespread. Final allocation of output was achieved through relatively decentralized, unplanned contracting. Although in theory prices were legally set from above, in practice the actual prices were often negotiated, and informal horizontal links were widespread.

Comparison between USSR and US Economies (1989) According to 1990 CIA World Factbook

	USSR	US
GDP (1989)	US$2.6595 trillion	US$5.2333 trillion
Population (July 1990)	290,938,469	250,410,000
GDP per capita	US$9,211	US$21,082
Labour force (1989)	152,300,000	125,557,000

A number of basic services were state-funded, such as education and healthcare. In the manufacturing sector, heavy industry and defense were assigned higher priority than consumer goods production. Consumer goods, in particular outside large cities, were often in short supply, of poor quality and limited choice, as under command economy consumers' preferences wielded almost no influence over production, changing demands of the population with growing money incomes couldn't be matched by supplies at rigidly fixed prices. A massive unplanned second economy existed alongside the planned one at low levels, providing some of the goods and services that the planners could not. Legalization of some elements of the decentralized economy was attempted with the reform of 1965.

Although statistics of the Soviet economy is notoriously unreliable and its growth is difficult to estimate precisely, by most accounts it continued to expand positively until 1989-1990. During 1950s and 1960s the Soviet economy performed with comparatively high growth rates and was catching up with the West. However, after 1970 the growth, while still positive, steadily declined, much more quickly and consistently than in other countries, despite rapid increase in the capital stock, surpassed only by that in Japan. In 1987 Mikhail Gorbachev pushed to reform the economy with his program of Perestroika in an attempt to revitalize it. His policies relaxed state control over enterprises, but hadn't yet allowed it to be replaced with market incentives, ultimately resulting in a sharp decline in production output. The economy, already suffering from reduced petroleum export revenues, started to collapse. Prices were still fixed, property was still largely state-owned until after the dissolution of the Soviet Union. For the most of the period after World War II and up to its collapse, the Soviet economy was the second largest in the world by GDP (PPP), though in per capita terms the Soviet GDP was behind that of the First World countries.

Culture

The culture of the Soviet Union passed through several stages during the USSR's 70-year existence. During the first eleven years following the Revolution (1918–1929), there was relative freedom and artists experimented with several different styles in an effort to find a distinctive Soviet style of art. Lenin wanted art to be accessible to the Russian people.

The government encouraged a variety of trends. In art and literature, numerous schools, some traditional and others radically experimental, proliferated. Communist writers Maksim Gorky and Vladimir Mayakovsky were active during this time. Film, as a means of influencing a largely illiterate society, received encouragement from the state; much of director Sergei Eisenstein's best work dates from this period.

Later, during Joseph Stalin's rule, Soviet culture was characterised by the rise and domination of the government-imposed style of Socialist realism, with all other trends being severely repressed, with rare exceptions (e.g. Mikhail Bulgakov's works). Many writers were imprisoned and killed. Also, religious people were persecuted and either sent to Gulags or were murdered by the thousands. The ban on the Orthodox Church was temporarily lifted in the 1940s, in order to rally support for the Soviet war against the invading forces of Germany. Under Stalin, prominent symbols that were not in line with communist ideology were destroyed, such as Orthodox Churches and Tsarist buildings.

Following the Khrushchev Thaw of the late 1950s and early 1960s, censorship was diminished. Greater experimentation in art forms became permissible once again, with the result that more sophisticated and subtly critical work began to be produced. The regime loosened its emphasis on socialist realism; thus, for instance, many protagonists of the novels of author Yury Trifonov concerned themselves with problems of daily life rather than with building socialism. An underground dissident literature, known as samizdat, developed during this late period. In architecture the Khrushchev era mostly focused on functional design as opposed to the highly decorated style of Stalin's epoch.

In the second half of the 1980s, Gorbachev's policies of perestroika and glasnost significantly expanded freedom of expression in the media and press, eventually resulting in the complete abolishment of censorship, total freedom of expression and freedom to criticise the government.

Russia

Russia, also officially known as the Russian Federation, is a country in northern Eurasia. It is a federal semi-presidential republic, comprising 83 federal subjects. Russia shares borders with the following countries (from northwest to southeast): Norway, Finland, Estonia, Latvia, Lithuania and Poland (both via Kaliningrad Oblast), Belarus, Ukraine, Georgia, Azerbaijan, Kazakhstan, China, Mongolia, and North Korea. It also has maritime borders with Japan (by the Sea of Okhotsk) and the United States (by the Bering Strait).

At 17,075,400 square kilometres (6,592,800 sq mi), Russia is by far the largest country in the world, covering more than a ninth of the Earth's land area. Russia is also the ninth most populous nation in the world with 142 million people. It extends across the whole of northern Asia and 40% of Europe, spanning 9 time zones and incorporating a wide range of environments and landforms. Russia has the world's largest reserves of mineral and energy resources, and is considered an energy superpower. It has the world's largest forest reserves and its lakes contain approximately one-quarter of the world's fresh water.

The nation's history began with that of the East Slavs, who emerged as a recognizable group in Europe between the 3rd and 8th centuries AD. Founded and ruled by a noble Viking warrior class and their descendants, the first East Slavic state, Kievan Rus', arose in the 9th century and adopted Orthodox Christianity from the Byzantine Empire in 988, beginning the synthesis of Byzantine and Slavic cultures that defined Russian culture for the next millennium. Kievan Rus' ultimately disintegrated and the lands were divided into many small feudal states. The most powerful successor state to Kievan Rus' was Moscow, which served as the main force in the Russian reunification process and independence struggle against the Golden Horde. Moscow gradually reunified the surrounding Russian principalities and came to dominate the cultural and political legacy of Kievan Rus'. By the 18th century, the nation had greatly expanded through conquest, annexation, and exploration to become the Russian Empire, which was the third largest empire in history, stretching from Poland in Europe to Alaska in North America.

Russia established worldwide power and influence from the times of the Russian Empire to being the largest and leading constituent of the Soviet Union, the world's first constitutionally socialist state and a recognized superpower, that played a decisive role in the allied victory in World War II. The Soviet era saw some of the greatest technology achievements of the nation, such as the world's first human spaceflight. The Russian Federation was founded following the dissolution of the Soviet Union in 1991, but is recognized as the continuing legal personality of the Soviet state. Russia has the world's 12th largest economy by nominal GDP or the seventh largest by purchasing power parity, with the fifth largest nominal military budget. It is one of the five recognized nuclear weapons states and possesses the world's largest stockpile of weapons of mass destruction. Russia is a great power and a permanent member of the United Nations Security Council, a member of the G8, G20, the Council of Europe, the Asia-Pacific Economic Cooperation, the Shanghai Cooperation Organization, the Eurasian Economic Community, the OSCE, and is the leading member of the Commonwealth of Independent States.

Etymology

The country´s original name was Donu (Rus'), a medieval state populated mostly by the Early East Slavs. However, this proper name became more prominent in the later history, and the country typically was called by its inhabitants "Donueay Caiey" (ruskaya zemlya) which could be translated as "Russian Land" or "Land of Rus'". The difference between the two terms would be like the difference between, for example, "Germany" and "German Land" or "Land of Germans". In order to distinguish this state from other states derived from it, it is denoted as Kievan Rus' by modern historiography.

An old Latin version of the name Rus' was Ruthenia, mostly applied to the western and southern regions of Rus' that were adjacent to Catholic Europe. The current name of the country, Dinney (Rossiya), comes from the Greek version of Rus', spelled Nuoba [roseia], which was the denomination of Kievan Rus in the Byzantine Empire.

Geography

Russia is the largest country in the world; its total area is 17,075,400 square kilometres (6,592,800 sq mi). The country contains 23 UNESCO World Heritage Sites, 40 UNESCO Biosphere reserves, 40 National Parks and 101 nature reserves. Russia has a wide natural resource base, including major deposits of timber, petroleum, natural gas, coal, ores and other mineral resources.

Topography

The two widest separated points in Russia are about 8,000 km (4,971 mi) apart along a geodesic line. These points are: the boundary with Poland on a 60 km (37 mi) long spit of land separating the Gulf of Gdansk from the Vistula Lagoon;

and the farthest southeast of the Kuril Islands, a few miles off Hokkaido Island, Japan. The points which are furthest separated in longitude are 6,600 km (4,101 mi) apart along a geodesic. These points are: in the west, the same spit; in the east, the Big Diomede Island (Ostrov Ratmanova). The Russian Federation spans 9 time zones. With access to three of the world's oceans — the Atlantic, Arctic, and Pacific — Russian fishing fleets are a major contributor to the world's fish supply. The Caspian is the source of what is considered one of the finest caviar in the world.

Most of Russia consists of vast stretches of plains that are predominantly steppe to the south and heavily forested to the north, with tundra along the northern coast. Russia possesses 10% of the world's arable land. Mountain ranges are found along the southern borders, such as the Caucasus (containing Mount Elbrus, which at 5,642 m (18,510 ft) is the highest point in both Russia and Europe) and the Altai (containing Mount Belukha, which at the 4,506 m (14,783 ft) is the highest point of Asian Russia); and in the eastern parts, such as the Verkhoyansk Range or the volcanoes on Kamchatka. The Ural Mountains, rich in mineral resources, form a north-south range that divides Europe and Asia. Russia has an extensive coastline of over 37,000 km (22,991 mi) along the Arctic and Pacific Oceans, as well as along the Baltic Sea, Sea of Azov, Black and Caspian seas.

The Barents Sea, White Sea, Kara Sea, Laptev Sea, East Siberian Sea, Chukchi Sea, Bering Sea, Sea of Okhotsk, and the Sea of Japan are linked to Russia via the Arctic and Pacific oceans. Russia's major islands and archipelagos include: Novaya Zemlya, the Franz Josef Land, the Severnaya Zemlya, the New Siberian Islands, Wrangel Island, the Kuril Islands, and Sakhalin. The Diomede Islands (one controlled by Russia, the other by the United States) are just 3 km (1.9 mi) apart, and Kunashir Island is about 20 km (12.4 mi) from Hokkaido.

Russia has thousands of rivers and inland bodies of water providing it with one of the world's largest surface water resources. The largest and most prominent of Russia's bodies of fresh water is Lake Baikal, the world's deepest, purest, oldest and most capacious freshwater lake. Lake Baikal alone contains over one fifth of the world's fresh surface water. Other major lakes include Lake Ladoga and Lake Onega, two of the largest lakes in Europe. Russia is second only to Brazil in volume of total renewable water resources. Of the country's 100,000 rivers, the Volga is the most famous, not only because it is the longest river in Europe, but also because of its major role in Russian history.

Climate

The climate of the Russian Federation formed under the influence of several determining factors. The enormous size of the country and the remoteness of many areas from the sea result in the dominance of the humid continental and subarctic climate, which is prevalent in European and Asian Russia except for the tundra and the extreme southeast. Mountains in the south obstruct the flow of warm air masses from the Indian Ocean, while the plain of the west and north makes the country open to Arctic and Atlantic influences.

Throughout much of the territory there are only two distinct seasons — winter and summer; spring and autumn are usually brief periods of change between extremely low temperatures and extremely high. The coldest month is January (February on the shores of the sea), the warmest usually is July. Great ranges of temperature are typical. In winter, temperatures get colder both from south to north and from west to east. Summers can be quite hot, even in Siberia. A small part of Black Sea coast around Sochi has a subtropical climate. The continental interiors are the driest areas.

Flora and Fauna

From north to south the East European Plain, also known as Russian Plain, is clad sequentially in Arctic *tundra*, coniferous forest (*taiga*), mixed and broad-leaf forests, grassland (*steppe*), and semi-desert (fringing the Caspian Sea), as the changes in vegetation reflect the changes in climate. Siberia supports a similar sequence but largely is taiga. Russia has the world's largest forest reserves, known as *"the lungs of Europe"*, second only to the Amazon Rainforest in the amount of carbon dioxide it absorbs.

There are 266 mammal species and 780 bird species in Russia. A total of 415 animal species have been included in the Red Data Book of the Russian Federation as of 1997 and are now protected.

Russian Federation

During and after the disintegration of the USSR, when wide-ranging reforms including privatisation and market and trade liberalization were being undertaken, the Russian economy went through a major crisis. The period was characterized by deep contraction of output, with GDP declining by roughly 50% between 1990 and the end of 1995 and industrial output declining by over 50%.

In October 1991, Yeltsin announced that Russia would proceed with radical, market-oriented reform along the lines of "shock therapy", as recommended by the United States and International Monetary Fund. Price controls were abolished, privatization was started. Millions plunged into poverty, from 1.5% of the population living in poverty in the late Soviet era, to 39%–49% by mid-1993.

Delays in wage payment became a chronic problem with millions being paid months, even years late. Russia took up the responsibility for settling the USSR's external debts, even though its population made up just half of the population of the USSR at the time of its dissolution. The privatization process largely shifted control of enterprises from state agencies to groups of individuals with inside connections in the Government and the mafia. Corruption became an everyday rule of life. Many of the newly rich mobsters and businesspeople took billions in cash and assets outside of the country in an enormous capital flight. The depression of state and economy led to the collapse of social services; the birth rate plummeted while the death rate skyrocketed. The early and mid-1990s saw extreme lawlessness, rise of criminal gangs and violent crime.

The 1990s were plagued by armed conflicts in the Northern Caucasus, both ethnic conflicts between local groups and separatist Islamist insurrections against federal power. Since the Chechen separatists had declared independence in the early 1990s, an intermittent guerrilla war was fought between the rebel groups and the Russian military. Terrorist attacks against civilians carried out by separatists, most notably the Moscow theater hostage crisis and Beslan school siege, caused hundreds of deaths and drew worldwide attention.

High budget deficits and the 1997 Asian Financial Crisis caused the financial crisis of 1998 and resulted in further GDP decline. On 31 December 1999 President Yeltsin resigned, handing the post to the recently appointed Prime Minister, Vladimir Putin, who then won the 2000 presidential election.

Putin suppressed the Chechen insurgency, although sporadic violence still occurs throughout the Northern Caucasus. High oil prices and initially weak currency followed by increasing domestic demand, consumption and investments has helped the economy grow for nine straight years, improving the standard of living and increasing Russia's influence on the world stage. While many reforms made during the Putin presidency have been generally criticized by Western nations as un-democratic, Putin's leadership over the return of order, stability, and progress has won him widespread popularity in Russia. On 2 March 2008, Dmitry Medvedev was elected President of Russia, whilst Putin became Prime Minister.

Government and Politics

According to the Constitution, which was adopted by national referendum on 12 December 1993 following the 1993 Russian constitutional crisis, Russia is a federation and formally a semi-presidential republic, wherein the President is the head of state and the Prime Minister is the head of government. The Russian Federation is fundamentally structured as a representative democracy. Executive power is exercised by the government.

Legislative power is vested in the two chambers of the Federal Assembly. The government is regulated by a system of checks and balances defined by the Constitution of the Russian Federation, which serves as the country's supreme legal document and as a social contract for the people of the Russian Federation. The federal government is composed of three branches:

- Legislative: The bicameral Federal Assembly, made up of the State Duma and the Federation Council adopts federal law, declares war, approves treaties, has the power of the purse, and has power of impeachment, by which it can remove the President.
- Executive: The president is the commander-in-chief of the military, can veto legislative bills before they become law, and appoints the Cabinet and other officers, who administer and enforce federal laws and policies.
- Judiciary: The Constitutional Court, Supreme Court, Supreme Court of Arbitration and lower federal courts, whose judges are appointed by the Federation Council on the recommendation of the president, interpret laws and can overturn laws they deem unconstitutional.

According to the Constitution, the justice in the court is based on the equality of all citizens, judges are independent and subject only to the law, trials are to be open and the accused is guaranteed a defense. Since 1996, Russia has instituted a moratorium on the death penalty, although capital punishment has not been abolished by law.

The president is elected by popular vote for a six-year term (eligible for a second term but constitutionally barred for a third consecutive term); election last held in 2008. Ministries of the government are composed of the premier and his deputies, ministers, and selected other individuals; all are appointed by the president on the recommendation of the Prime Minister (whereas the appointment of the latter requires the consent of the State Duma). The national legislature is the Federal Assembly, which consists of two chambers; the 450-member State Duma and the 176-member Federation Council. Leading political parties in Russia include United Russia, the Communist Party, the Liberal Democratic Party of Russia, and Fair Russia.

Human Rights

The rights and liberties of the citizens of the Russian Federation are granted by Chapter 2 of the Constitution. Russia is a signatory to the Universal Declaration of Human Rights and has also ratified a number of other international human rights instruments.

In 2004, Alvaro Gil-Robles, the first Commissioner for Human Rights of the Council of Europe, said that *"the fledgling Russian democracy is still, of course, far from perfect, but its existence and its successes cannot be denied."*

However, some leading international democracy and human rights organizations consider Russia to have not enough democratic attributes and to allow few political rights and civil liberties to its citizens. US-funded international organization Freedom House ranks Russia as "not free", citing "carefully engineered elections" and "absence" of debate. Amnesty International accuses Russia of committing wide ranging human rights abuses, including granting impunity for murderers of human rights activists, imprisoning political dissidents and operating a system of arbitrary arrest. Human Rights Watch claims Russia commits grave human rights violations in Chechnya and allows the systematic abuse of migrant workers. Press freedom in Russia is considered amongst the lowest in the world by press freedom organization Reporters Without Borders and is ranked 141st in their annual survey, on the basis that the Russian authorities "black list" figures that are critical of the government, practice "official harassment", and "gag" potential dissidents.

Russian authorities dismiss these claims and especially criticise Freedom House. The Ministry of Foreign Affairs of Russia has called the 2006 Freedom in the World Report "prefabricated"; the ministry also claims that such organizations as Freedom House and Human Rights Watch use the same scheme of voluntary extrapolation of *"isolated facts that of course can be found in any country"* into *"dominant tendencies"*. The chairwoman of the Civil Society Institution and Human Rights Council at the President of Russia Ella Pamfilova also criticized the Freedom House views on Russia as *"ridiculous, absurd and far-fetched"*.

Foreign Relations

The Russian Federation is recognized in international law as successor state of the former Soviet Union. Russia continues to implement the international commitments of the USSR, and has assumed the USSR's permanent seat on the UN Security

Council, membership in other international organizations, the rights and obligations under international treaties and property and debts. Russia has a multifaceted foreign policy. As of 2009, it maintains diplomatic relations with 191 countries and has 144 embassies. The foreign policy is determined by the President of Russia and implemented by the Ministry of Foreign Affairs.

As one of five permanent members of the UN Security Council, Russia plays a major role in maintaining international peace and security. The country participates in the Quartet on the Middle East and the Six-party talks with North Korea. Russia is a member of the Group of Eight (G8) industrialized nations, the Council of Europe, OSCE and APEC. Russia usually takes a leading role in regional organizations such as the CIS, Eurasec, CSTO, and the SCO. Former President Vladimir Putin had advocated a strategic partnership with close integration in various dimensions including establishment of four common spaces between Russia and the EU. Since the collapse of the Soviet Union, Russia has developed a friendlier, albeit volatile relationship with NATO. The NATO-Russia Council was established in 2002 to allow the 26 Allies and Russia to work together as equal partners to pursue opportunities for joint collaboration.

Military

Russia assumed control of Soviet assets abroad and most of the Soviet Union's production facilities and defense industries. The Russian military is divided into the Ground Forces, Navy, and Air Force. There are also three independent arms of service: Strategic Rocket Forces, Military Space Forces, and the Airborne Troops. In 2006, the military had 1.037 million personnel on active duty.

Russia has the largest stockpile of nuclear weapons in the world. It has the second largest fleet of ballistic missile submarines and is the only country apart from the U.S. with a modern strategic bomber force. Russia's tank force is the largest in the world, its surface navy and air force are among the strongest.

The country has a large and fully indigenous arms industry, producing most of its own military equipment with only few types of weapons imported. Russia is the world's top supplier of arms, a spot it has held since 2001, accounting for around 30% of worldwide weapons sales and exporting weapons to about 80 countries.

It is mandatory for all male citizens aged 18–27 to be drafted for a year of service in Armed Forces; the government plans to increase the proportion of contract servicemen to 70% by 2010. Defense expenditure has quadrupled over the past six years. According to Stockholm International Peace Research Institute estimates, official government military spending for 2008 was $58 billion, the fifth largest in the world, though various sources, including US intelligence, and the International Institute for Strategic Studies, have estimated Russia's military expenditures to be considerably higher. Currently, the military is undergoing a major equipment upgrade worth about $200 billion between 2006 and 2015. Defense Minister Anatoliy Serdyukov supervises the major reforms aimed to transform a mass mobilization army into a smaller force of contract soldiers.

Subdivisions

The Russian Federation comprises 83 federal subjects. These subjects have equal representation—two delegates each—in the Federation Council. However, they differ in the degree of autonomy they enjoy.

- 46 oblasts (provinces): most common type of federal subjects, with federally appointed governor and locally elected legislature.
- 21 republics: nominally autonomous; each has its own constitution, president, and parliament. Republics are allowed to establish their own official language alongside Russian but are represented by the federal government in international affairs. Republics are meant to be home to specific ethnic minorities.
- 9 krais (territories): essentially the same as oblasts. The "territory" designation is historic, originally given to frontier regions and later also to administrative divisions that comprised autonomous okrugs or autonomous oblasts.

- 4 autonomous okrugs (autonomous districts): originally autonomous entities within oblasts and krais created for ethnic minorities, their status was elevated to that of federal subjects in the 1990s. With the exception of Chukotka Autonomous Okrug, all autonomous okrugs are still administratively subordinated to a krai or an oblast of which they are a part.
- 1 autonomous oblast (the Jewish Autonomous Oblast): originally autonomous oblasts were administrative units subordinated to krais. In 1990, all of them except the Jewish AO were elevated in status to that of a republic.
- 2 federal cities (Moscow and St. Petersburg): major cities that function as separate regions.

Federal Districts

Federal subjects are grouped into 8 federal districts, each administered by an envoy appointed by the President of Russia. Unlike the federal subjects, the federal districts are not a subnational level of government, but are a level of administration of the federal government. Federal districts' envoys serve as liaisons between the federal subjects and the federal government and are primarily responsible for overseeing the compliance of the federal subjects with the federal laws.

Demographics

Ethnic Composition (2002) Russians	**79.8%**
Tatars	3.8%
Ukrainians	2.0%
Bashkirs	1.2%
Chuvash	1.1%
Chechen	0.9%
Armenians	0.8%
Other/unspecified	10.4%

Ethnic Russians comprise 79.8% of the population; however the Russian Federation is also home to several sizeable minorities. In total, 160 different other ethnic groups and indigenous peoples live within its borders. Though Russia's population is comparatively large, its density is low because of the country's enormous size. Population is densest in European Russia, near the Ural Mountains, and in southwest Siberia. 73% of the population lives in urban areas while 27% in rural ones. The population of Russia is 141,927,297 as of 1 January 2010.

In 2008, the population declined by 121,400 people, or by "0.085% (in 2007 – by 212,000, or 0.15% and in 2006 – by 532,600 people, or 0.37%). In 2008 migration continued to grow by a pace of 2.7% with 281,615 migrants arriving to the Russian Federation, of which 95% came from CIS countries, the vast majority being Russians or Russian speakers.

The number of Russian emigrants declined by 16% to 39,508, of which 66% went to other CIS countries. There are also an estimated 10 million illegal immigrants from the ex-Soviet states in Russia. Roughly 116 million ethnic Russians live in Russia and about 20 million more live in other former republics of the Soviet Union, mostly in Ukraine and Kazakhstan.

The population of Russia peaked at 148,689,000 in 1991, just before the breakup of the Soviet Union. It began to experience a rapid decline starting in the mid-90s. The decline has slowed to near stagnation in recent years due to reduced death rates, increased birth rates and increased immigration. The number of deaths during 2008 was 363,500 greater than the number of births. This is down from 477,700 in 2007, and 687,100 in 2006. According to data published by the Russian Federal State Statistics Service, the mortality rate in Russia declined 4% in 2007, as compared to 2006, reaching some 2 million deaths, while the birth rate grew 8.3% year-on-year to an estimated 1.6 million live births.

The primary causes of Russia's population decrease are a high death rate and low birth rate. While Russia's birth-rate is comparable to that of other European countries (12.1 births per 1000 people in 2008 compared to the European Union average of 9.90 per 1000) its population is declining at a greater rate than many due to a substantially higher death rate (in 2008, Russia's death rate was 14.5 per 1000 people compared to the European Union average of 10.28 per 1000). However, the Russian Ministry of Health and Social Affairs predicts that by 2011, the death rate will equal the birth rate due to increases in fertility and decline in mortality.

Russia's 160 ethnic groups speak some 100 languages. According to the 2002 census, 142.6 million people speak Russian, followed by Tatar with 5.3 million and Ukrainian with 1.8 million speakers. Russian is the only official state language, but the Constitution gives the individual republics the right to make their native language co-official next to Russian.

Despite its wide dispersal, the Russian language is homogeneous throughout Russia. Russian is the most geographically widespread language of Eurasia and the most widely spoken Slavic language. Russian belongs to the Indo-European language family and is one of the living members of the East Slavic languages; the others being Belarusian and Ukrainian (and possibly Rusyn). Written examples of Old East Slavic (*Old Russian*) are attested from the 10th century onwards.

Over a quarter of the world's scientific literature is published in Russian. Russian is also applied as a means of coding and storage of universal knowledge—60–70% of all world information is published in the English and Russian languages. The language is one of the six official languages of the United Nations.

Religion

Christianity, Islam, Buddhism and Judaism are Russia's traditional religions, deemed part of Russia's "historical heritage" in a law passed in 1997. Estimates of believers widely fluctuate among sources, and some reports put the number of non-believers in Russia at 16–48% of the population. Russian Orthodoxy is the dominant religion in Russia. 95% of the registered Orthodox parishes belong to the Russian Orthodox Church while there are a number of smaller Orthodox Churches. However, the vast majority of Orthodox believers do not attend church on a regular basis. Nonetheless, the church is widely respected by both believers and nonbelievers, who see it as a symbol of Russian heritage and culture. Smaller Christian denominations such as Roman Catholics, Armenian Gregorians, and various Protestants exist. The ancestors of many of today's Russians adopted Orthodox Christianity in the 10th century.

The 2007 International Religious Freedom Report published by the US Department of State said that approximately 100 million citizens consider themselves Russian Orthodox Christians. According to a poll by the Russian Public Opinion Research Center, 63% of respondents considered themselves Russian Orthodox, 6% of respondents considered themselves Muslim and less than 1% considered themselves either Buddhist, Catholic, Protestant or Jewish. Another 12% said they believe in God, but did not practice any religion, and 16% said they are non-believers.

It is estimated that Russia is home to some 15–20 million Muslims. However, the Islamic scholar and human rights activist Roman Silantyev has claimed that there are only 7 to 9 million people who adhere to the Islamic religion in Russia. Russia also has an estimated 3 million to 4 million Muslim migrants from the ex-Soviet states. Most Muslims live in the Volga-Ural region, as well as in the North Caucasus, Moscow, Saint Petersburg and western Siberia.

Buddhism is traditional for three regions of the Russian Federation: Buryatia, Tuva, and Kalmykia. Some residents of the Siberian and Far Eastern regions, Yakutia, Chukotka, etc., practice shamanist, pantheistic, and pagan rites, along with the major religions. Induction into religion takes place primarily along ethnic lines. Slavs are overwhelmingly Orthodox Christian. Turkic speakers are predominantly Muslim, although several Turkic groups in Russia are not.

Health

The Russian Constitution guarantees free, universal health care for all citizens. In practice, however, free health care is partially restricted due to propiska regime. While Russia has more physicians, hospitals, and health care workers than almost any other country in the world on a per capita basis, since the collapse of the Soviet Union the health of the Russian population has declined considerably as a result of social, economic, and lifestyle changes. As of 2007, the average life expectancy in Russia is 61.5 years for males and 73.9 years for females. The combined average Russian life expectancy of 67.7 years at birth is 10.8 years shorter than the overall figure in the European

Union. The biggest factor contributing to this relatively low life expectancy for males is a high mortality rate among working-age males from preventable causes (e.g., alcohol poisoning, stress, smoking, traffic accidents, violent crimes). Mortality among Russian men rose by 60% since 1991, four to five times higher than in Europe. As a result of the large difference in life expectancy between men and women and because of the lasting effect of World War II, where Russia lost more men than any other nation in the world, the gender imbalance remains to this day and there are 0.859 males to every female.

Heart diseases account for 56.7% of total deaths, with about 30% involving people still of working age. A study blamed alcohol for more than half the deaths (52%) among Russians aged 15 to 54 from 1990 to 2001. For the same demographic, this compares to 4% of deaths for the rest of the world. About 16 million Russians suffer from cardiovascular diseases, placing Russia second in the world, after Ukraine, in this respect. Death rates from homicide, suicide, and cancer are also especially high. 52% of men and 15% of women smoke, more than 260,000 lives believed to be lost each year as a result of tobacco use.

HIV/AIDS, virtually non-existent in the Soviet era, rapidly spread following the collapse, mainly through the explosive growth of intravenous drug use. According to official statistics, there are currently more than 364,000 people in Russia registered with HIV, but independent experts place the number significantly higher. In increasing efforts to combat the disease, the government increased spending on HIV control measures 20-fold in 2006, and the 2007 budget doubled that of 2006. Since the Soviet collapse, there has also been a dramatic rise in both cases of and deaths from tuberculosis, with the disease being particularly widespread amongst prison inmates.

In an effort to stem Russia's demographic crisis, the government is implementing a number of programs designed to increase the birth rate and attract more migrants to alleviate the problem. The government has doubled monthly child support payments and offered a one-time payment of 250,000 Rubles (around US$10,000) to women who had a second child since 2007. In 2007, Russia saw the highest birth rate since the collapse of the USSR. The First Deputy PM also said about 20 billion rubles (about US$1 billion) will be invested in new prenatal centres in Russia in 2008–2009. Immigration is increasingly seen by the United Nations as necessary to sustain the country's population.

Education

Russia has a free education system guaranteed to all citizens by the Constitution, and has a literacy rate of 99.4%. Entry to higher education is highly competitive. As a result of great emphasis on science and technology in education, Russian medical, mathematical, scientific, and space and aviation research is generally of a high order.

Before 1990 the course of school training in Soviet Union was 10-years, but at the end of 1990 the 11-year course has been officially entered. Education in state-owned secondary schools is free; *first* tertiary (university level) education is free with reservations: a substantial share of students is enrolled for full pay (many state institutions started to open commercial positions in the last years). In 2004 state spending for education amounted to 3.6% of GDP, or 13% of consolidated state budget.

The Government allocates funding to pay the tuition fees within an established quota, or number of students for each state institution. This is considered crucial because it provides access to higher education to all skilled students, as opposed to only those who can afford it. In addition, students are paid a small stipend and provided with free housing. Apart from state higher education institutions, many private ones have emerged to address the need for a skilled work-force for high-tech and emerging industries and economic sectors.

Economy

The economic crisis that struck all post-Soviet countries in the 1990s was nearly twice as intense as the Great Depression in the countries of Western Europe and the United States in the 1930s. Even before the financial crisis of 1998, Russia's GDP was half of what it had been in the early 1990s. Since the turn of the century, rising oil prices, increased foreign investment, higher domestic consumption and greater political stability have bolstered economic growth in Russia. The country ended 2007 with its ninth straight year of growth, averaging 7% annually since 1998. In 2007, Russia's GDP was $2.076 trillion (est. PPP), the 6th largest in the world, with GDP growing 8.1% from the previous year. Growth was primarily driven by non-traded services and goods for the domestic market, as opposed to oil or mineral extraction and exports.

The average salary in Russia was $640 per month in early 2008, up from $80 in 2000. Approximately 14% of Russians lived below the national poverty line in 2007, significantly down from 40% in 1998 at the worst of the post-Soviet collapse. Unemployment in Russia was at 6% in 2007, down from about 12.4% in 1999.

Oil, natural gas, metals, and timber account for more than 80% of Russian exports abroad. Since 2003, however, exports of natural resources started decreasing in economic importance as the internal market strengthened considerably. Despite higher energy prices, oil and gas only contribute to 5.7% of Russia's GDP and the government predicts this will drop to 3.7% by 2011. Russia is also considered well ahead of most other resource-rich countries in its economic development, with a long tradition of education, science, and industry. The country has more higher education graduates than any other country in Europe.

A simpler, more streamlined tax code adopted in 2001 reduced the tax burden on people, and dramatically increased state revenue. Russia has a flat personal income tax rate of 13 percent. This ranks it as the country with the second most attractive personal tax system for single managers in the world after the United Arab Emirates.

The federal budget has run surpluses since 2001 and ended 2007 with a surplus of 6% of GDP. Over the past several years, Russia has used oil revenues from its Stabilization Fund of the Russian Federation to prepay most of its formerly massive debts, leaving it with one of the lowest foreign debts among major economies.

Oil export earnings have allowed Russia to increase its foreign reserves from $12 billion in 1999 to $597.3 billion on

1 August 2008, the third largest reserves in the world. The economic development of the country though has been uneven geographically with the Moscow region contributing a disproportionately high amount of the country's GDP. Much of Russia, especially indigenous and rural communities in Siberia, lags significantly behind. Nevertheless, the middle class has grown from just 8 million persons in 2000 to 55 million persons in 2006. Over the last five years, fixed capital investments have averaged real gains greater than 10% per year and personal incomes have achieved real gains more than 12% per year.

Despite the country's strong economic performance since 1999, however, the World Bank lists several challenges facing the Russian economy including its diversification, encouraging the growth of small and medium enterprises, building human capital and improving corporate governance. Another problem is modernisation of infrastructure, ageing and inadequate after years of being neglected; the government has said $1 trillion will be invested in development of infrastructure by 2020.

Agriculture

The total area of cultivated land in Russia was estimated as 1,237,294 km in 2005, the fourth largest in the world. Unlike most other countries, Russia has large reserves of unused arable land, in part due to the drop in agricultural production during the economy crisis of 1990s, when the area planted to grains dropped by 25%. This was accompanied by a severe decline of livestock inventories.

In 1999–2009, however, Russia's agriculture demonstrated steady growth, and the country turned from a grain importer to the third largest grain exporter after EU and U. S. in 2009. The production of meat has grown from 6,813,000 tonnes in 1999 to 9,331,000 tonnes in 2008, and continues to grow.

This restoration of agriculture was supported by successful farm credit policy of the government, helping both individual farmers and large privatized corporate farms, that once were Soviet kolkhozes and still own the significant share of agricultural land. While large individual farms and corporate farms concentrate mainly on the production of grain (including for export), as well as husbandry products, small private household plots produce most of the country's yield of potatoes, vegetables and fruits.

Energy

Russia is known as an energy superpower. The country has the world's largest natural gas reserves, the 8th largest oil reserves, and the second largest coal reserves. Russia is the world's leading natural gas exporter and leading natural gas producer, while also the second largest oil exporter and largest oil producer, though Russia interchanges the latter status with Saudi Arabia from time to time.

Russia is the 4th largest electricity generator in the world and the 5th largest renewable energy producer, the latter due to the well-developed hydroelectricity production in the country. Large cascades of hydropower plants are built in European Russia along big rivers like Volga. The Asian part of Russia also features a number of major hydropower stations, however the gigantic hydroelectric potential of Siberia and the Russian Far East largely remains unexploited.

Russia was the first country to develop civilian nuclear reactor and to introduce the first nuclear power plant. Currently, Russia is the 4th largest nuclear energy producer. Rosatom Nuclear Energy State Corporation manages all the nuclear plants in Russia. Nuclear energy is rapidly developing in Russia, with the aim of increasing the total share of nuclear energy from current 16.9% to 23% by 2020. The Russian government plans to allocate 127 billion rubles ($5.42 billion) to a federal program dedicated to the next generation of nuclear energy technology. About 1 trillion rubles ($42.7 billion) is to be allocated from the federal budget to nuclear power and industry development before 2015. Russia remains among the world leaders in nuclear technology and is a member of ITER international fusion reactor project.

Science and Technology

At the start of the 18th century the reforms of Peter the Great (the founder of Russian Academy of Sciences and Saint Petersburg State University) and the work of such champions as polymath Mikhail Lomonosov (the founder of Moscow State

University) gave a great boost for development of science and innovation in Russia.

In the 19th and 20th centuries the country produced a large number of great scientists and inventors.

In mathematics Nikolai Lobachevsky, a *Copernicus of Geometry*, founded the non-Euclidean geometry. At least after such prominent scientist as Chebyshev the Russian mathematical school became one of the most influential ones in the world and was represented by numerous figures greatly contributing to different fields of mathematics, physics and computing sciences. Chebyshev's students included Aleksandr Lyapunov who founded the modern stability theory (lately deeply developed by such scientists as Aleksandr Andronov and Vladimir Arnold), and Andrey Markov who developed the theory of Markov chains, playing a central role in information sciences and modern applied mathematics. Nikolai Zhukovsky was a founding father of the modern aero-and hydrodynamics. Andrei Kolmogorov, a leading mathematician of the 20th century, developed the modern theory of probability and made other major contributions to broadest range of mathematical branches, including turbulence, topology and information theory. Israel Gelfand is credited with many important discoveries in mathematical physics and applied sciences. Sergei Sobolev developed a theory of Sobolev space which played an extremely important role in formation of modern mathematical views and introduced a notion of distributions generalyzing ideas of Newton and Leibniz. Yevgraf Fyodorov was a founder of modern crystallography. Nicolai Vasiliev was a forerunner of paraconsistent and multi-valued logics. Such mathematicians as Pavel Alexandrov, Lev Pontryagin, Andrey Tychonoff, and many others made fundamental contributions to different fields of mathematics. Nine Soviet/Russian mathematicians were awarded with Fields Medal, a most prestigious award in mathematics. Recently Grigori Perelman was offered the first ever Clay Millennium Prize Problems award for his final proof of the Poincare conjecture in 2002.

In chemistry Dmitry Mendeleev invented the Periodic table, that is the main framework of the modern chemistry, while Aleksandr Butlerov was one of the creators of the theory of chemical structure, playing a central role in organic chemistry. Nikolay Semyonov made major contributions to explanation of the mechanism of chemical transformation (1956 Nobel Prize in Chemistry).

The Russian physics school began to develop after Lomonosov. During the period of origin of electrodynamics Vasily Petrov discovered the electric arc effect in 1802 and Heinrich Lenz discovered an important law named in his honour. Nikolay Umov discovered a fundamental concept of Umov-Poynting vector and was the first scientist to indicate interrelation between mass and energy proposing the formula $E = kmc$ as early as in 1873. Alexander Popov was among the inventors of radio. Russian/Soviet physics in the 20th century was one of leading ones in the world. Nikolay Bogolyubov suggested a triplet quark model, introduced a new quantum degree of freedom (later called as colour charge) for quarks and formulated a microscopic theory of superconductivity. Lev Landau made fundamental contributions to many areas of theoretical physics. Nikolai Basov and Alexander Prokhorov were co-inventors of lasers and masers. Igor Tamm, Andrei Sakharov and Lev Artsimovich developed the idea of tokamak for controlled nuclear fusion and created its first prototype, which finally led to the modern ITER project. Yevgeny Zavoisky discovered electron paramagnetic resonance playing important role in studying chemical species.

In biology Dmitry Ivanovsky was the first scientist to discover viruses (1892). Ivan Pavlov is widely known for first describing the phenomenon of classical conditioning. Ilya Mechnikov was a pioneer in investigations of the immune system (1908, Nobel Prize in Medicine).

In computing sciences the first ternary computer Setun was developed by Nikolay Brusentsov, together with Sergei Sobolev, in 1958 and Sergei Lebedev developed one of the first universally programmable computers in continental Europe in 1950, MESM.

Nikolay Benardos introduced the arc welding, further developed by Nikolay Slavyanov, Konstantin Khrenov and other Russian engineers. Gleb Kotelnikov invented the knapsack parachute, while Evgeniy Chertovsky introduced the pressure suit. Pavel Yablochkov and Alexander Lodygin were pioneers of electric street lighting, and Mikhail Dolivo-Dobrovolsky invented the three-phase electric power system, widely used today.

Many famous Russian scientists and inventors were emigres, like Igor Sikorsky, credited with invention of first helicopters, and Vladimir Zworykin, often called the father of TV, chemist Ilya Prigogine, noted for his work on dissipative structures and complex systems (1977 Nobel Prize for Chemistry), economists Simon Kuznets (1971 Nobel Prize) and Wassily Leontief (1973 Nobel Prize), physicist Georgiy Gamov (an author of the Big Bang theory) and social scientist Pitirim Sorokin who played an important role in development of sociology in the USA. Many foreigners worked in Russia for a long time, like Leonard Euler and Alfred Nobel.

The greatest Russian successes are in the field of space technology and space exploration. Konstantin Tsiolkovsky was the father of theoretical austronautics. His works had inspired leading Soviet rocket engineers such as Sergey Korolyov, Valentin Glushko and many others that contributed to the success of the Soviet space program at early stages of the Space Race and beyond.

In 1957 the first Earth-orbiting artificial satellite, *Sputnik 1*, was launched; in 1961 on 12 April the first human trip into space was successfully made by Yury Gagarin; and many other Soviet and Russian space exploration records ensued, including the first spacewalk performed by Alexey Leonov, the first space exploration rover *Lunokhod-1* and the first space station *Salyut 1*. Nowadays Russia is the largest satellite launcher and the only provider of transport for space tourism services.

Other technologies, where Russia historically leads, include nuclear technology, aircraft production and arms industry.

The creation of the first nuclear power plant along with the first nuclear reactors for submarines and surface ships was directed by Igor Kurchatov. NS *Lenin* was the world's first nuclear powered surface ship as well as the first nuclear powered civilian vessel, and NS *Arktika* became the first surface ship to reach the North Pole. The history of the Russian aircraft eingineering originated from a pioneer of aviation Alexander Mozhaysky who made his first attempt to fly by his own design aircraft (monoplane) as early as in 1881. In the 20th century a number of prominent Soviet aerospace engineers, inspired by the theoretical works of Nikolai Zhukovsky, supervised the creation of many dozens of models of military and civilian aircraft and founded a number of *KBs* (*Construction Bureaus*) that now constitute the bulk of Russian United Aircraft Corporation.

Famous Russian airplanes include the first supersonic passenger jet Tupolev Tu-144 by Alexei Tupolev, MiG fighter aircraft series by Artem Mikoyan and Mikhail Gurevich, and Su series by Pavel Sukhoi and his followers. MiG-15 is the world's most produced jet aircraft in history, while MiG-21 is the most produced supersonic aircraft. During World War II era Bereznyak-Isayev BI-1 was introduced as the first rocket-powered fighter aircraft, and Ilyushin Il-2 bomber became the most produced military aircraft in history. Polikarpov Po-2 *Kukuruznik* is the world's most produced biplane, and Mil Mi-8 is the most produced helicopter.

Famous Russian battle tanks include T-34, the best tank design of World War II, and further tanks of T-series, including the most produced tank in history, T-54/55, the first fully gas turbine tank T-80 and the most modern Russian tank T-90. The AK-47 and AK-74 by Mikhail Kalashnikov constitute the most widely used type of assault rifle throughout the world — so much so that more AK-type rifles have been manufactured than all other assault rifles combined. With these and other weapons Russia for a long time has been among the world's top suppliers of arms, accounting for around 30% of worldwide weapons sales and exporting weapons to about 80 countries.

With such technological achievements, however, since the time of Brezhnev stagnation Russia was lagging significantly behind the West in a number of technologies, especially those concerning energy conservation and consumer goods production. The crisis of 1990-s led to the drastic reduction of the state support for science. Many Russian scientists and university graduates left Russia for Europe or United States; this migration is known as a brain drain. In 2000-s, on the wave of a new economic boom, the situation in the Russian science and technology has improved, and the government launched a campaign aimed into modernisation and innovation. Russia's President Dmitry Medvedev formulated top 5 priorities for the country's technological development: energy efficiency, IT (including both common products and the products combined with space technology), nuclear energy and pharmaceuticals. Some progress already has been achieved, with Russia's having nearly completed GLONASS, the only global satellite navigation system apart from American GPS, and Russia's being the only country constructing mobile nuclear plants.

Transportation

Railway transport in Russia is mostly under the control of the state-run Russian Railways monopoly. The company accounts for over 3.6% of Russia's GDP and handles 39% of the total of Russia's freight traffic (including pipelines) and more than 42% of passenger traffic. The total length of common-used railway tracks exceeeds 85,500 km, second only to the United States. Over 44,000 km of tracks are electrified, which is the largest number in the world, and additionally there are more than 30,000 km of industrial non-common carrier lines. Railways in Russia, unlike in the most of the world, use broad gauge of 1,520 mm (4 ft 11+D $_6$ in), with the exception of 957 km on Sakhalin Island using narrow gauge of 1,067 mm (3 ft 6 in). The most renown railroad in Russia is Trans-Siberian Railway or Transsib, spanning a record 7 time zones and serving the longest single continuous services in the world, Moscow-Vladivostok (9,259 km, 5,753 mi), Moscow–Pyongyang (10,267 km, 6,380 mi) and Kiev–Vladivostok (11,085 km, 6,888 mi).

As of 2006 Russia had 933,000 km of roads, of which 755,000 were paved. Some of these make up the Russian federal motorway system. With a large land area the road density is the lowest of all the G8 and BRIC countries. A Russian saying states that *There are two main problems in Russia: fools and roads*, however this very lack of roads was of much help to Russians in the times of Napoleon's and Hitler's invasions

102,000 km of inland waterways in Russia mostly go by natural rivers or lakes. In the European part of the country the network of channels connects the basins of major rivers. Russia's capital, Moscow, is sometimes called *"the port of the five seas"*, due to its waterway connections to the Baltic, White, Caspian, Azov and Black seas.

Major sea ports of Russia include Rostov-on-Don on the Azov Sea, Novorossiysk on the Black Sea, Astrakhan and Makhachkala on the Caspian Sea, Kaliningrad and St. Petersburg on the Baltic Sea, Arkhangelsk on the White Sea, Murmansk on the Barents Sea, Petropavlovsk-Kamchatsky and Vladivostok on the Pacific Ocean. In 2008 Russia owned 1448 merchant marine ships. Russia is the only country to have nuclear icebreaker fleet, which is a great advantage in the economic exploitation of Arctic continental shelf of Russia and the development of sea trade through the Northern Sea Route between Europe and East Asia.

There are 74,285 km of oil pipelines in Russia, 13,658 km of pipelines for refined products, 158,767 km of natural gas pipelines By total length of pipelines Russia is second only to the United States. Currently, many new pipeline projects are being realized, including Nord Stream and South Stream natural gas pipelines to Europe, and ESPO oil pipeline to Russian Far East and China.

Russia has 1216 airports, the busiest being Sheremetyevo, Domodedovo, and Vnukovo in Moscow and Pulkovo in Saint Petersburg. The total length of airlines in Russia exceeds 600,000 km. In the remote regions of the Russian North and Siberia the transportation by air (usually by helicopters) is vital, and in some months of the year it is the only transport link to the rest of the country.

Typically, major Russian cities have well-developed and diverse systems of public transport, with the most common varieties of exploited vehicles being bus, trolleybus and tram. Seven Russian cities, namely Moscow, Saint Petersburg, Nizhny Novgorod, Novosibirsk, Samara, Yekaterinburg and Kazan, have undeground metros, while Volgograd features a metrotram. Total length of metros in Russia is 465.4 km. Moscow Metro and Saint Petersburg Metro are the oldest in Russia, opened in 1935 and 1955 respectively. These two are among the fastest and busiest metro systems in the world, and are famous for rich decorations and unique designs of their stations, which is a common tradition for Russian metros and railways.

Culture

Folk Culture and Cuisine

There are over 160 different ethnic groups and indigenous peoples in Russia. Ethnic Russians with their Slavic Orthodox culture, Tatars and Bashkirs with their Turkic Muslim culture, Buddhist nomadic Buryats and Kalmyks, Shamanistic peoples of the Far North and Siberia, highlanders of the Northern Caucasus, Finno-Ugric peoples of the Russian North West and Volga Region all contribute to diverse and rich culture of Russia. The ethnic culture is preserved in various museums and ethnoparks, reproduced in cuisine, architecture, cinema and arts, and developed by folk bands, dance ensembles and choirs.

Woodcraft Russian architecture, widely associated with the ethnic culture, is at best represented in wooden churches. Russian traditional wooden dwelling is izba, while the early type of fortified settlements is known as kremlin. Handicraft, like Dymkovo toy, khokhloma, gzhel, Zhostovo painting, pisanka and palekh, is also associated with the folk culture. Ethnic Russian clothes include kaftan, kosovorotka and ushanka for men, sarafan and kokoshnik for women, with lapti and valenki as common shoes. The Cossacks of Southern Russia have a separate brand of culture within ethnic Russian, their clothes including burka and papaha, which they share with the peoples of the Northern Caucasus. Russian cuisine widely uses fish, poultry, mushrooms, berries, and honey. Crops of rye, wheat, barley, and millet provide the ingredients for a plethora of breads, pancakes, cereals, kvass, beer, and vodka. Black bread is relatively more popular in Russia if compared with the rest of the world. Flavourful soups and stews include shchi, borsch, ukha, solyanka and okroshka.

Smetana (a heavy sour cream) is often added to soups and salads. Pirozhki, blini and syrniki are native types of pancakes.

Cutlets (like Chicken Kiev), pelmeni and shashlyk are popular meat dishes, the last two being of Tatar and Caucasus origin respectively. Popular salads include Russian salad, vinaigrette and Dressed Herring.

Russians have many traditions, most prominent being the washing in banya, a hot steam bath somewhat similar to sauna. Old Russian folklore takes its roots in the pagan beliefs of ancient Slavs and now is represented in the Russian fairy tales. Epic Russian bylinas are another important part of Slavic mythology. The oldest bylinas of Kievan cycle were actually recorded mostly in the Russian North, especially in Karelia, where most of the Finnish national epic Kalevala was recorded as well. Russia's large number of ethnic groups have distinctive traditions of folk music. Typical ethnic Russian musical instruments are gusli, balalaika, zhaleika and garmoshka. Folk music had great influence on the Russian classical composers, and in modern times it is a source of inspiration for a number of popular folk bands, most prominent being Melnitsa. Russian folk songs, as well as patriotic songs of the Soviet era, constitute the bulk of repertoire of the world-renown Red Army choir and other popular Russian ensembles.

Many Russian fairy tales and bylinas were adaptated for animation films, or for feature movies by the prominent directors like Aleksandr Ptushko (*Ilya Muromets*, *Sadko*) and Aleksandr Rou (*Morozko*, *Vasilisa the Beautiful*). Some Russian poets, including Pyotr Yershov and Leonid Filatov, made a number of well-known poetical interpretations of the classical Russian fairy tales, and in some cases, like that of Alexander Pushkin, also created fully original fairy tale poems of great popularity.

Architecture

Russian architecture began with the woodcraft buildings of ancient Slavs. Since Christianization of Kievan Rus' for several ages Russian architecture was influenced predominantly by the Byzantine architecture, until the Fall of Constantinople. Apart from fortifications (kremlins), the main stone buildings of ancient Rus' were Orthodox churches, with their many domes, often gilded or brightly painted. Aristotle Fioravanti and other Italian architects brought Renaissance trends into Russia.

The 16th century saw the development of unique tent-like churches culminating in Saint Basil's Cathedral. By that time the onion dome design was also fully developed. In the 17th century, the "fiery style" of ornamentation flourished in Moscow and Yaroslavl, gradually paving the way for the Naryshkin baroque of the 1690s. After Peter the Great reforms had made Russia much closer to Western culture, the change of the architectural styles in Russia generally followed that of Western Europe.

The 18th-century taste for rococo architecture led to the splendid works of Bartolomeo Rastrelli and his followers. During the reign of Catherine the Great and her grandson Alexander I, the city of Saint Petersburg was transformed into an outdoor museum of Neoclassical architecture.

The second half of the 19th century was dominated by the Byzantine and Russian Revival style (this corresponds to Gothic Revival in Western Europe). Prevalent styles of the 20th century were the Art Nouveau (Fyodor Shekhtel), Constructivism (Aleksey Shchusev and Konstantin Melnikov), and the Stalin Empire style (Boris Iofan).

After Stalin's death a new Soviet leader, Nikita Khruschev, condemned the "excesses" of the former architectural styles, and in the late Soviet era the architecture of the country was dominated by plain functionalism. This helped somewhat to resolve the housing problem, but created a large quantity of buildings of low architectural quality, much in contrast with the previous bright architecture. After the end of the Soviet Union the situation improved. Many churches demolished in Soviet times were rebuilt, and this process continues along with the restoration of various historical buildings destroyed in World War II. As for the original architecture, there is no longer any common style in modern Russia, though International style has a great influence.

Visual Arts

Early Russian painting focused on icon painting and vibrant frescos inherited by Russians from Byzantium. As Moscow rose to power, Theophanes the Greek and Andrei Rublev became vital names associated with the beginning of a distinctly Russian art. The Russian Academy of Arts was created in 1757, aimed to give Russian artists an international role and status. Notable portrait painters from the Academy include Ivan Argunov, Fyodor Rokotov, Dmitry Levitzky, and Vladimir Borovikovsky. In the early 19th century, when neoclassicism and romantism flourished, famous academic artists focused on mythological and Biblical themes, like Karl Briullov and Alexander Ivanov.

Realism came into dominance in the 19th century. The realists captured Russian identity in landscapes of wide rivers, forests, and birch clearings, as well as vigorous genre scenes and robust portraits of their contemporaries. Other artists focused on social criticism, showing the conditions of the poor and caricaturing authority; critical realism flourished under the reign of Alexander II, with some artists making the circle of human suffering their main theme. Others focused on depicting dramatic moments in Russian history.

The *Peredvizhniki* (*wanderers*) group of artists broke with Russian Academy and initiated a school of art liberated from Academic restrictions. Leading realists include Ivan Shishkin, Arkhip Kuindzhi, Ivan Kramskoi, Vasily Polenov, Isaac Levitan, Vasily Surikov, Viktor Vasnetsov, and Ilya Repin.

By the turn of the 20th century and on, many Russian artists developed their own vividly unique styles, neither realist nor avantegarde. These include Boris Kustodiev, Kuzma Petrov-Vodkin, Mikhail Vrubel and Nicholas Roerich.

The Russian avant-garde is an umbrella term used to define the large, influential wave of modernist art that flourished in Russia from approximately 1890 to 1930. The term covers many separate, but inextricably related, art movements that occurred at the time; namely neo-primitivism, suprematism, constructivism, rayonism, and futurism. Notable artists from this era include El Lissitzky, Kazimir Malevich, Wassily Kandinsky, Vladimir Tatlin, Alexander Rodchenko, and Marc Chagall. The Russian avant-garde reached its creative and popular height in the period between the Russian Revolution of 1917 and 1932, at which point the revolutionary ideas of the avant-garde clashed with the newly emerged conservative direction of socialist realism.

In the Soviet era many artists combined innovation with socialist realism including Ernst Neizvestny, Ilya Kabakov, Mikhail Shemyakin, Erik Bulatov, and Vera Mukhina. They employed techniques as varied as primitivism, hyperrealism, grotesque, and abstraction. Soviet artists produced works that were furiously patriotic and anti-fascist in the 1940s. After the Great Patriotic War Soviet sculptors made multiple monuments to the war dead, marked by a great restrained solemnity.

In the 20th century many Russian artists made their careers in Western Europe, forced to emigrate by the Revolution. Wassily Kandinsky, Marc Chagall, Naum Gabo and others spread their work, ideas, and the impact of Russian art globally.

Classical Music and Ballet

Music in 19th century Russia was defined by the tension between classical composer Mikhail Glinka along with his followers, who embraced Russian national identity and added religious and folk elements to their compositions, and the Russian Musical Society led by composers Anton and Nikolay Rubinstein, which was musically conservative. The later Romantic tradition of Tchaikovsky, one of the greatest composers of the Romantic era, whose music has come to be known and loved for its distinctly Russian character as well as its rich harmonies and stirring melodies, was brought into the 20th century by Sergei Rachmaninoff, one of the last great champions of the Romantic style of European classical music.

World-renowned composers of the 20th century included Scriabin, Stravinsky, Rachmaninoff, Prokofiev, Shostakovich and Sviridov. During most of the Soviet Era, music was highly scrutinized and kept within a conservative, accessible idiom in conformity with the policy of socialist realism.

Soviet and Russian conservatories have turned out generations of world-renowned soloists. Among the best known are violinists David Oistrakh and Gidon Kremer; cellist Mstislav Rostropovich; pianists Vladimir Horowitz, Sviatoslav Richter, and Emil Gilels; and vocalists Fyodor Shalyapin, Galina Vishnevskaya, Anna Netrebko and Dmitry Hvorostovsky. During the early 20th century, Russian ballet dancers Anna Pavlova and Vaslav Nijinsky rose to fame, and impresario

Sergei Diaghilev and his Ballets Russes' travels abroad profoundly influenced the development of dance worldwide. Soviet ballet preserved the perfected 19th century traditions, and the Soviet Union's choreography schools produced one internationally famous star after another, including Maya Plisetskaya, Rudolf Nureyev, and Mikhail Baryshnikov. The Bolshoi Ballet in Moscow and the Mariinsky in Saint Petersburg remain famous throughout the world.

Literature and Philosophy

Russian literature is considered to be among the most influential and developed in the world, contributing many of the world's most famous literary works. Russia's literary history dates back to the 10th century; in the 18th century its development was boosted by the works of Mikhail Lomonosov and Denis Fonvizin, and by the early 19th century a modern native tradition had emerged, producing some of the greatest writers of all time. This period and the Golden Age of Russian Poetry began with Alexander Pushkin, considered to be the founder of modern Russian literature and often described as the *"Russian Shakespeare"*.

It continued in the 19th century with the poetry of Mikhail Lermontov and Nikolay Nekrasov, dramas of Aleksandr Ostrovsky and Anton Chekhov, and the prose of Nikolai Gogol, Ivan Turgenev, Leo Tolstoy, Fyodor Dostoevsky, Mikhail Saltykov-Shchedrin, Ivan Goncharov, Aleksey Pisemsky and Nikolai Leskov. Tolstoy and Dostoevsky in particular were titanic figures to the point that many literary critics have described one or the other as the greatest novelist ever.

By the 1880s Russian literature had begun to change. The age of the great novelists was over and short fiction and poetry became the dominant genres of Russian literature for the next several decades which became known as the Silver Age of Russian Poetry. Previously dominated by realism, Russian literature came under strong influence of Symbolism in the years between 1893 and 1914. Leading writers of this age include Valery Bryusov, Andrei Bely, Vyacheslav Ivanov, Aleksandr Blok, Nikolay Gumilev, Dmitry Merezhkovsky, Fyodor Sologub, Anna Akhmatova, Osip Mandelstam, Marina

Tsvetaeva, Leonid Andreyev, Ivan Bunin, and Maxim Gorky. Some Russian writers, like Tolstoy and Dostoyevsky, are known also as philosophers, while many more authors are known primarily for their philosophical works. Russian philosophy blossomed since the 19th century, when it was defined initially by the opposition of Westernizers, advocating Russia's following the Western political and economical models, and Slavophiles, insisting on developing Russia as unique civilization.

The latter group includes Nikolai Danilevsky and Konstantin Leontiev, the early founders of eurasianism. In its further development, Russian philosophy was always marked by deep connection to literature and interest in creativity, society, politics and nationalism; cosmos and religion were other primary subjects. Notable philosopheres of the late 19th and early 20th centuries include Vladimir Solovyev, Sergei Bulgakov, Pavel Florensky and Vladimir Vernadsky. In the 20th century Russian philosophy became dominated by Marxism. Following the Russian Revolution of 1917 and the ensuing civil war, Russian cultural life was left in chaos. Some prominent writers and philosophers, like Ivan Bunin, Vladimir Nabokov, Lev Shestov, Isaiah Berlin, Alexandre Kojeve left the country, while a new generation of talented writers joined together in different organizations with the aim of creating a new and distinctive working-class culture appropriate for the new state, the Soviet Union.

Throughout the 1920s writers enjoyed broad tolerance. In the 1930s censorship over literature was tightened in line with Joseph Stalin's policy of socialist realism. After his death the restrictions on literature were eased, and by the 1970s and 1980s, writers were increasingly ignoring the official guidelines. The leading authors of the Soviet era included Yevgeny Zamiatin, Isaac Babel, Vladimir Mayakovsky, Ilf and Petrov, Yury Olesha, Mikhail Bulgakov, Boris Pasternak, Mikhail Sholokhov, Aleksandr Solzhenitsyn, Yevgeny Yevtushenko, and Andrey Voznesensky.

Cinema, Animation and Media

While in the industrialized nations of the West, motion pictures had first been accepted as a form of cheap recreation and leisure for the working class, Russian filmmaking came to prominence following the 1917 revolution when it explored editing as the primary mode of cinematic expression. Russian and later Soviet cinema was a hotbed of invention in the period immediately following the 1917, resulting in world-renowned films such as *Battleship Potemkin*. Soviet-era filmmakers, most notably Sergei Eisenstein and Andrei Tarkovsky, would become some of the world's most innovative and influential directors.

Eisenstein was a student of filmmaker and theorist Lev Kuleshov, who developed the groundbreaking Soviet montage theory of film editing at the world's first film school, the All-Union Institute of Cinematography. Dziga Vertov, whose *kino-glaz* ("film-eye") theory—that the camera, like the human eye, is best used to explore real life—had a huge impact on the development of documentary film making and cinema realism. In 1932, Stalin made socialist realism the state policy; this somewhat limited creativity, however many Soviet films in this style were artistically successful, like *Chapaev*, *The Cranes Are Flying*, and *Ballad of a Soldier*.

1960s and 1970s saw a greater variety of artistic styles in the Soviet cinema. Eldar Ryazanov's and Leonid Gaidai's comedies of that time were immensely popular, with many of the catch phrases still in use today. In 1961–1968 Sergey Bondarchuk directed an Oscar-winning film adaptation of Tolstoy's epic *War and Peace*, which was the most expensive film ever made. In 1969, Vladimir Motyl's *White Sun of the Desert* was released, a very popular film in a genre known as 'osterns'; the film is traditionally watched by cosmonauts before any trip into space.

Russia also has a long and rich tradition of animation, which started already in the late Russian Empire times. Most of Russia's cartoon production for cinema and television was created during Soviet times, when Soyuzmultfilm studio was the largest animation producer. Soviet animators developed a great and unmatched variety of pioneering techniques and aesthetic styles, with prominent directors including Ivan Ivanov-Vano, Fyodor Khitruk and Aleksandr Tatarskiy. Soviet cartoons are still a source for many popular catch phrases, while such cartoon heroes as Russian-style Winnie-the-Pooh, cute little

Cheburashka, Wolf and Hare from *Nu, Pogodi!* being iconic images in Russia and many surrounding countries.

The late 1980s and 1990s were a period of crisis in Russian cinema and animation. Although Russian filmmakers became free to express themselves, state subsidies were drastically reduced, resulting in fewer films produced. The early years of the 21st century have brought increased viewership and subsequent prosperity to the industry on the back of the economy's rapid development, and production levels are already higher than in Britain and Germany. Russia's total box-office revenue in 2007 was $565 million, up 37% from the previous year (by comparison, in 1996 revenues stood at $6 million). Russian cinema continues to receive international recognition. *Russian Ark* (2002) was the first feature film ever to be shot in a single take. The traditions of Soviet animation were developed in the past decade by such directors as Aleksandr Petrov and studios like Melnitsa.

Russia was among the first countries to introduce radio and television. Due to the enormous size of the country Russia leads in the number of TV broadcast stations and repeaters. There were few channels in the Soviet time, but in the past two decades many new state-run and private-owned radio stations and TV channels appeared. In 2005 a state-run English language Russia Today TV started broadcasting, and its Arabic version Rusiya Al-Yaum was launched in 2007.

Modern Culture

Since the late Soviet times Russia has experienced another wave of Western cultural influence, which led to the development of many previously unknown phenomena in the Russian culture. Russia easily has adopted a number of cultural techniques, while providing its own content.

The most vivid example, perhaps, is the Russian rock music, which takes its roots both in the Western rock and roll and heavy metal, and in traditions of the Russian bards of Soviet era, like Vladimir Vysotsky and Bulat Okudzhava. Saint-Petersburg (former Leningrad), Yekaterinburg and Omsk became the main centres of development of the rock music. Popular Russian rock groups include Mashina Vremeni, DDT,

Aquarium, Alisa, Kino, Nautilus Pompilius, Aria, Grazhdanskaya Oborona, Splean and Korol i Shut.

At the same time Russian pop music developed from what was known in the Soviet times as *estrada* into full-fledged industry, with some performers gaining international recognition, like t.A.T.u. in the West or Vitas in China. Lubeh is a very popular and unique group, harmoniously combining the elements of Western rock and roll, traditional Russian folk music and military bard music, featuring a number of rock attributes but often performing on the pop scenes.

In the past decades many new sporting activities came into Russia, including cheerleading, auto racing, snowboarding and skateboarding. Many subcultures became popular among Russian youth, like rappers, Goths, Emo, Anime fans and Live action role-playing gamers. Russian Internet, or Runet, has seen a rapid development in the last years and the rise of a variety of Internet subcultures.

Sports

Russians have been successful at a number of sports and consistently finish in the top rankings at the Olympic Games and in other international competitions. Combining the total medals of Soviet Union and Russia, the country is second among all nations by number of gold medals both at the Summer Olympics and at the Winter Olympics.

During the Soviet era, the national Olympic team placed first in the total number of medals won at 14 of its 18 appearances; with these performances, the USSR was the dominant Olympic power of its era. Since the 1952 Olympic Games, Soviet and later Russian athletes have always been in the top three for the number of gold medals collected at the Summer Olympics.

Soviet gymnasts, track-and-field athletes, weight lifters, wrestlers, boxers, fencers, shooters, chess players, cross country skiers, biathletes, speed skaters and figure skaters were consistently among the best in the world, along with Soviet basketball, handball, volleyball and ice hockey players. Since the collapse of the Soviet Union, Russian athletes have continued to dominate international competitions. The 1980 Summer Olympic Games were held in Moscow while the 2014 Winter Olympics will be hosted by Sochi.

As the Soviet Union, Russia was traditionally very strong in basketball, winning various Olympic tournaments, World Championships and Eurobasket. As of 2009 they have various players in the NBA, notably Utah Jazz forward Andrei Kirilenko, and are considered as a worldwide basketball force. In 2007, Russia defeated world champions Spain to win Eurobasket 2007. Russian basketball clubs such as PBC CSKA Moscow (2006 and 2008 Euroleague Champions) have also had great success in European competitions such as the Euroleague and the ULEB Cup.

Although ice hockey was only introduced during the Soviet era, the national team soon dominated the sport internationally, winning gold at almost all the Olympics and World Championships they contested. Russian players Valery Kharlamov, Sergey Makarov, Vyacheslav Fetisov and Vladislav Tretiak hold 4 of 6 positions in the IIHF *Team of the Century*. As with some other sports, the Russian ice hockey programme suffered after the breakup of the Soviet Union with Russia enduring a 15 year gold medal drought. At that time many prominent Russian players made their career in the NHL.

In recent years Russia has reemerged as a hockey superpower, winning back to back gold medals in the 2008 and 2009 World Championships, and overtaking team Canada as the top ranked ice hockey team in the world. The KHL (Kontinental Hockey League) was founded in 2008 as a successor to the Russian Superleague. It is seen as a rival to the NHL and is ranked the top hockey league in Europe as of 2009. Bandy, known in Russian as "hockey with a ball", is another traditionally popular ice sport, with national league games averaging around 3500 spectators. The Soviet Union won all the Bandy World Championships from 1957 to 1979.

During the Soviet period, Russia was also a competitive footballing nation. Despite having fantastic players, the USSR never really managed to assert itself as one of the major forces of international football, although its teams won various championships (such as Euro 1960) and reached numerous finals (such as Euro 1988). Along with ice hockey and basketball, football is one of the most popular sports in modern Russia. In recent years, Russian football, which downgraded in 1990-s, has experienced a revival. Russian clubs (such as CSKA Moscow, Zenit St Petersburg, Lokomotiv Moscow, and Spartak Moscow) are becoming increasingly successful on the European stage (CSKA and Zenit winning the UEFA Cup in 2005 and 2008 respectively). The Russian national football team reached the semi-finals of Euro 2008, losing only to eventual champions Spain.

Soviet Union dominated the sport of gymnastics for many years, with such athletes as Larisa Latynina, who currently holds a record of most Olympic medals won per person and most gold Olympic medals won by a woman. Today, Russia is leading in rhythmic gymnastics with such stars as Alina Kabayeva, Irina Tschaschina and Yevgeniya Kanayeva. Russian synchronized swimming is the best in the world, with almost all gold medals having been swept by Russians at Olympics and World Championships for more than a decade.

Figure skating is another popular sport in Russia; in the 1960s, the Soviet Union rose to become a dominant power in figure skating, especially in pair skating and ice dancing, and at every Winter Olympics from 1964 until 2006, a Soviet or Russian pair has won gold, often considered the longest winning streak in modern sports history. Since the end of the Soviet era, tennis has grown in popularity and Russia has produced a number of famous tennis players. Chess is also a widely popular pastime; from 1927, Soviet and Russian chess grandmasters have held the world championship almost continuously.

National Holidays and Symbols

There are seven public holidays in Russia. The New Year is the first in calendar and in popularity. Russian New Year traditions resemble those of the Western Christmas, with New Year Trees and gifts, and Ded Moroz (Father Frost) playing the same role as Santa. Rozhdestvo (Orthodox Christmas) falls on 7 January, because Russian Orthodox Church still follows the Julian (old style) calendar and all Orthodox holidays are 13 days after Catholic ones. Another two major Christian holidays are Paskha (Easter) and Troitsa (Trinity), but there is no need to recognize them as public holidays since they are always celebrated on Sunday. Kurban Bayram and Uraza Bayram are widely celebrated by Russian Muslims.

Further Russian public holidays include Defender of the Fatherland Day (23 February), which honors Russian men, especially those serving in the army; International Women's Day (8 March), which combines the traditions of Mother's Day and Valentine's Day; International Workers' Day (1 May), now renamed *Spring and Labour Day*; Victory Day (9 May); Russia Day (12 June); and Unity Day (4 November), commemorating the popular uprising which expelled the Polish-Lithuanian occupation force from Moscow in 1612. The latter is a replacement for the old Soviet holiday celebrating October Revolution of 1917 (again, it was falling on November because of the difference of calendars). Fireworks and outdoor concerts are common features of all Russian public holidays.

Victory Day is the second popular holiday in Russia, it commemorates the victory over Nazi Germany in World War II and is widely celebrated throughout the country. A huge military parade, hosted by the President of the Russian Federation, is annually organized in Moscow on Red Square. Similar parades are organized in all major Russian cities and the cities with the status *Hero city* or *City of Military Glory*.

Other popular holidays, which are not public, include Old New Year (New Year according to Julian Calendar on 14 January), Tatiana Day (day of Russian students on 25 January), Maslenitsa (an old pagan holiday a week before the Great Lent), Cosmonautics Day (a day of Yury Gagarin's first ever human trip into space on 12 April), Ivan Kupala Day (another pagan Slavic holiday on 7 July) and Peter and Fevronia Day (taking place on 8 July and being the Russian analogue of Valentine's Day, which focuses, however, on the family love and fidelity). On different days in June there are major celebrations of the end of the school year, when graduates from schools and universities traditionally swim in the city fountains; the local varieties of these public events include Scarlet Sails tradition in Saint Petersburg. State symbols of Russia include the Byzantine double-headed eagle, combined with St. George of Moscow in the Russian coat of arms; these symbols date from the Grand Duchy of Moscow time. Russian flag appeared in the late Tsardom of Russia period and became widely used since Russian Empire times. Russian anthem shares its music with the Soviet Anthem, though not the lyrics (many Russians of older generations just don't know the new lyrics and sing the old ones).

The Russian imperial motto *God is with us* and the Soviet motto *Proletarians of all countries, unite!* are now obsolete and no new motto has been officially introduced to replace them. The hammer and sickle and the full Soviet coat of arms are still widely seen in Russian cities as a part of old architectural decorations. The Soviet Red Stars are also encountered, often on military equipment and war memorials. The Red Banner continues to be honoured, especially the Banner of Victory of 1945.

Matryoshka doll is a recognizable symbol of Russia, while the towers of Moscow Kremlin and Saint Basil's Cathedral in Moscow are main Russia's architectural symbols. Cheburashka is a mascot of Russian national Olympic team. Mary, Saint Nicholas, Saint Andrew, Saint George, Saint Alexander Nevsky, Saint Sergius of Radonezh and Saint Seraphim of Sarov are Russia's patron saints.

Chamomile is a flower that Russians often associate with their Motherland, while birch is a national tree. Russian bear is an animal symbol and national personification of Russia, though this image has Western origin and Russians themselves have accepted it fairly recently. The native Russian national personification is Mother Russia, sometimes called Mother Motherland.

Tourism

Tourism in Russia has seen rapid growth since the late Soviet times, first inner tourism and then international tourism as well. Rich cultural heritage and great natural variety place Russia among the most popular tourist destinations in the world. The country contains 23 UNESCO World Heritage Sites, while many more are on UNESCO's tentative lists. Major tourist routes in Russia include a travel around the Golden Ring of ancient cities, cruises on the big rivers like Volga, and long journeys on the famous Trans-Siberian Railway.

Most popular tourist destinations in Russia are Moscow and Saint Petersburg, the current and the former capitals of the country and great cultural centres, recognized as World Cities. Moscow and Saint Petersburg feature such world-renown museums as Tretyakov Gallery and Hermitage, famous theaters like Bolshoi and Mariinsky, ornate churches like Saint Basil's Cathedral, Cathedral of Christ the Saviour, Saint Isaac's Cathedral and Church of the Savior on Blood, impressive fortifications like Moscow Kremlin and Peter and Paul Fortress, beautiful squares like Red Square and Palace Square, and busy streets like Tverskaya and Nevsky Prospect.

Rich palaces and parks of extreme beauty are found in the former imperial residences in suburbs of Moscow (Kolomenskoye, Tsaritsyno) and Saint Petersburg (Peterhof, Strelna, Oranienbaum, Gatchina, Pavlovsk, Tsarskoye Selo). Moscow contains a great variety of impressive Soviet era buildings along with modern scyscrapers, while Saint Petersburg, nicknamed *Venice of the North*, boasts of its classical architecture, many rivers, channels and bridges.

Kazan, the capital of Tatarstan, shows a unique mix of Christian Russian and Muslim Tatar cultures. The city has registered a brand *The Third Capital of Russia*, though a number of other major Russian cities compete for this status, like Novosibirsk, Yekaterinburg and Nizhny Novgorod, all being major cultural centres with rich history and prominent architecture. Veliky Novgorod, Pskov and the cities of Golden Ring (Vladimir, Yaroslavl, Kostroma and others) have at best preserved the architecture and the spirit of ancient and medieval Rus', and also are among the main tourist destinations. Many old fortifications (typically Kremlins), monasteries and churches are scattered throughout Russia, forming its unique cultural landscape both in big cities and in remote areas.

Typical Russian souvenirs include matryoshka doll and other handicraft, samovars for water heating, ushanka and papaha warm hats, fur clothes and other stuff. Russian vodka and caviar are among the food that attracts foreigners, along with honey, blini, pelmeni, borsch and other products and dishes. Diverse regions and ethnic cultures of Russia offer many more different food and souvenirs, and show a great variety of traditions, like Russian banya, Tatar Sabantuy, or Siberian shamanist rituals.

The warm subtropical Black Sea coast of Russia is the site for a number of popular sea resorts, like Sochi, known for its beaches and wonderful nature. The mountains of the Northern Caucasus contain popular ski resorts, including Dombay.

The most famous natural tourist destination in Russia is lake Baikal, named *the Blue Eye of Siberia*. This unique lake, oldest and deepest in the world, has crystal-clean waters and is surrounded by taiga-covered mountains. Other popular natural destinations include Kamchatka with its volcanoes and geysers, Karelia with its many lakes and granite rocks, Altai with its snowy mountains and Tyva with its wild steppes.

5

Central Europe and the Balkan Countries

The Balkans (often referred to as the Balkan Peninsula, although the two are not coterminous) is a geopolitical and cultural region of southeastern Europe. The region takes its name from the Balkan Mountains, which run through the centre of Bulgaria into eastern Serbia. The region has a combined area of 550,000 km (212,000 sq mi) and a population of 55 million people.

"*Balkan*" comes from a Turkish word meaning "a chain of wooded mountains". The ancient Greek name for the Balkan Peninsula was the "Peninsula of Haemus". Romans considered the Rhodope Mountains to be the northern limit of the Peninsula of Haemus and the same limit applied approximately to the border between Greek and Latin use in the region (this was later called Jireeek Line). The Balkans are also referred to as Southeastern Europe.

Definitions and Boundaries

The Balkan Peninsula

The Balkan Peninsula may be defined as an area of southeastern Europe surrounded by water on three sides: the Adriatic Sea to the west, the Mediterranean Sea (including the Ionian and Aegean seas) to the south and the Black Sea to the east. Its northern boundary is often given as the Danube, Sava and Kupa/Kolpa rivers.

Countries which are geographically fully located within the Balkan peninsula:

- Albania
- Bosnia and Herzegovina
- Bulgaria
- Greece
- Kosovo *(partially recognized)*
- Macedonia
- Montenegro.

Countries which are significantly located on the peninsula:

- Serbia (Central Serbia, partially Belgrade region; comprises around 2/3 of country's territory)
- Croatia (around 1/2 of overall territory)
- Slovenia (around 1/3 of overall territory).

Countries which are located mostly outside the peninsula:

- Romania (6% of its territory, comprising Northern Dobruja)
- Turkey (3% of its territory, comprising Eastern Thrace)
- Italy (small part around the city of Trieste).

The Balkans

The term "The Balkans" covers not only those countries which lie within the boundaries of the "Balkan Peninsula", but may also include Slovenia, and Romania. Slovenia, which was part of Yugoslavia from 1919 to 1991, lies partially north of the Danube-Sava line and therefore outside the Peninsula, but prior to 1991 the whole of Yugoslavia was considered to be part of the Balkans. The father of the term "The Balkans" August Zeune defined it in 1808 to describe areas that remained under Turkish rule after 1699.

In most of the English-speaking world, the countries commonly included in the Balkan region are:

- Albania
- Bosnia and Herzegovina

- Bulgaria
- Croatia
- Greece
- Kosovo *(partially recognized)*
- Macedonia
- Montenegro
- Serbia.

Other countries sometimes included are:

- Moldova
- Romania
- Slovenia
- Turkey.

Italy until WWII included Istria and some Dalmatian areas (like Zara), but now it has only the small area of the Province of Trieste inside the Balkan region (but it is usually considered not part of the Balkans by Italian geographers like Vittorio Vialli: for them even Istria is not in the Balkans, because they connect the Kupa river with Rijeka (Fiume) and consider this to be the Balkans western limit).

Etymology and Evolving Meaning

The region takes its name from the Stara Planina mountain range in Bulgaria, commonly known as the Balkan Mountains (from the Turkish *balkan* meaning "a chain of wooded mountains"). The name is still preserved in Central Asia where there exist the Balkan Mountains and the Balkan Province of Turkmenistan. On a larger scale, the mountains are only one part of a long continuous chain of mountains crossing the region in the form of a reversed letter S, from the Carpathians south to the Balkan range proper, before marching away east into Anatolian Turkey. On the west coast, an offshoot of the Dinaric Alps follows the coast south through Dalmatia and Albania, crosses Greece and continues into the sea in the form of various islands.

The first attested time the name "Balkan" was used in the West for the mountain range in Bulgaria was in a letter sent in 1490 to Pope Innocent VIII by Buonaccorsi Callimaco, an Italian humanist, writer and diplomat. English traveler John Morritt introduced this term into the English literature at the end of the 18th century, and other authors started applying the name to the wider area between the Adriatic and the Black Sea. The concept of the "Balkan peninsula" was created by the German geographer August Zeune in 1808.

As time passed, the term gradually obtained political connotations far from its initial geographic meaning, arising from political changes from the late 1800s to the creation of post-World War I Yugoslavia (initially the Kingdom of Serbs, Croats and Slovenes). Zeune's goal was to have a geographical parallel term to the Italic and Iberian Peninsula, and seemingly nothing more. The gradually acquired political connotations are newer, and, to a large extent, due to oscillating political circumstances.

The term Balkans is generally used to describe areas that remained under Turkish rule after 1699, namely: Bulgaria, Serbia (except for Vojvodina), Macedonia, Thrace, Albania, Valahia, Moldavia, Epirus, Bosnia and Herzegovina, Montenegro (except for the Boka Bay and Budva), central Greece and the Peloponnese. Vojvodina and Transylvania, it is argued, do not belong to Balkans. After the split of Yugoslavia beginning in June 1991, the term 'Balkans' again received a negative meaning, even in casual usage. Over the last decade, in the wake of the former Yugoslav split, many Slovenians and Croatians, as well as Serbs of Vojvodina have attempted to reject their label as 'Balkan nations'.

This is in part due to the pejorative connotation of the term 'Balkans' in the 1990s, and continuation of this meaning until now. Today, the term 'Southeast Europe' is preferred or, in the case of Slovenia and Croatia, 'Central Europe' and Greece has almost exclusively been regarded and referred to as a 'Southern European' country.

Southeastern Europe

Because of the negative connotations of the term "Balkan", writers such as Maria Todorova and Vesna Goldsworthy have suggested the use of the term Southeastern Europe instead.

The use of this term is slowly growing; a European Union initiative of 1999 is called the *Stability Pact for South Eastern Europe*, and the online newspaper *Balkan Times* renamed itself *Southeast European Times* in 2003.

Western Balkans

European Union institutions and member states define the "Western Balkans" as Albania and the former Yugoslavia, minus Slovenia. The European Bank for Reconstruction and Development uses "Western Balkans" to refer to the above states, minus Croatia. Today Western Balkans is more of a political than geographic definition for the region of Southeast Europe that is not yet in the European Union.

Nature and Natural Resources

Most of the area is covered by mountain ranges running from north-west to south-east. The main ranges are the Dinaric Alps in Slovenia, Croatia and Bosnia, the Sar massif which spreads from Albania to Macedonia and the Pindus range, spanning from southern Albania into central Greece. In Bulgaria there are ranges running from east to west: the Balkan mountains and the Rhodope mountains at the border with Greece. The highest mountain of the region is Rila in Bulgaria, with Musala at 2925 m, with Mount Olympus in Greece, the throne of Zeus, being second at 2919 m and Vihren in Bulgaria being the third at 2914 m.

On the coasts the climate is Mediterranean, in the inland it is moderate continental. In the northern part of the peninsula and on the mountains, winters are frosty and snowy, while summers are hot and dry. In the southern part winters are milder.

During the centuries many woods have been cut down and replaced with bush. In the southern part and on the coast there is evergreen vegetation. In the inland there are woods typical of Central Europe (oak and beech, and in the mountains, spruce, fir and pine). The tree line in the mountains lies at the height of 1800–2300 m.

The soils are generally poor, except on the plains where areas with natural grass, fertile soils and warm summers provide an opportunity for tillage. Elsewhere, land cultivation is mostly unsuccessful because of the mountains, hot summers and poor soils, although certain cultures such as olives and grapes flourish.

Resources of energy are scarce, except in Kosovo, where considerable coal, lead, zinc, chromium, silver deposits are located. Other deposits of coal, especially in Romania, Bulgaria, Serbia and Bosnia also exist. Lignite deposits are widespread in Greece. Petroleum is most notably present in Romania, although scarce reserves exist in Greece, Serbia, Albania and Croatia. Natural gas deposits are scarce. Hydropower stations are largely used in energetics.

Metal ores are more usual than other raw materials. Iron ore is rare but in some countries there is a considerable amount of copper, zinc, tin, chromite, manganese, magnesite and bauxite. Some metals are exported.

History and Geopolitical Significance

The Balkan region was the first area of Europe to experience the arrival of farming cultures in the Neolithic era. The practices of growing grain and raising livestock arrived in the Balkans from the Fertile Crescent by way of Anatolia, and spread west and north into Pannonia and Central Europe.

The identity of the Balkans is dominated by its geographical position; historically the area was known as a crossroads of various cultures. It has been a juncture between the Latin and Greek bodies of the Roman Empire, the destination of a massive influx of pagan Slavs, an area where Orthodox and Catholic Christianity met, as well as the meeting point between Islam and Christianity.

In pre-classical and classical antiquity, this region was home to Greeks, Illyrians, Paeonians, Thracians, Dacians and other ancient groups. Later the Roman Empire conquered most of the region and spread Roman culture and the Latin language but significant parts still remained under classical Greek influence. The Slavs arrived in the 6th century and began assimilating and displacing the older inhabitants of the northern and central Balkans. During the Middle Ages, the Balkans became the stage for a series of wars between the Byzantine, Bulgarian and Serbian Empires.

By the end of the 16th century, the Ottoman Empire had become the controlling force in the region after expanding from Anatolia through Thrace to the Balkans. Many people in the Balkans and Carpathians place their greatest folk heroes in the era of either the onslaught or the retreat of the Ottoman Empire. As examples, for Croats, Nikola Subis Zrinski and Petar Kruzis; for Greeks Constantine XI Palaiologos and Kolokotronis for Serbs, Milos Obiliæ and Tzar Lazar; for Albanians, Gjergj Kastrioti Skanderbeg; for ethnic Macedonians, Nikola Karev; for Bosniaks, Husein Gradaseeviæ; and for Bulgarians, Vasil Levski, Georgi Sava Rakovski and Hristo Botev. In the past several centuries, because of the frequent Ottoman wars in Europe fought in and around the Balkans, and the comparative Ottoman isolation from the mainstream of economic advance (reflecting the shift of Europe's commercial and political centre of gravity towards the Atlantic), the Balkans has been the least developed part of Europe. The population of the Balkans, according to one estimate, fell from a high of 8 millions in the late 16th century to only 3 million by the mid-eighteenth.

The Balkan nations gained independence in the 19th century (Greece in 1829, Serbia in 1833, Bulgaria and Montenegro in 1878). In 1912–1913 the First Balkan War broke out when the Christian nations, having large populations under Ottoman rule, united in an alliance against the Turks and after a five months war essentially terminated the Turks' presence in Europe after five centuries. Two months after the end of the common war, a Second Balkan War broke out when Bulgaria dissatisfied from its share, attacked its allies Serbia and Greece. The Serbs and the Greeks repelled them and after Greek army invaded Bulgaria following by a Romanian intervention, Bulgaria collapsed. The Turks used the favourable opportunity to recapture Eastern Thrace, establishing its new western borders that stand until today.

The First World War was sparked in the Balkans in 1914 when a Black hand revolutionary organization with predominately Serbian and pro Yugoslav oriented members assassinated in Bosnia and Herzegovina's capital Sarajevo the Austro-Hungarian heir Archduke Franz Ferdinand of Austria. That caused a war between the two countries which-through the existing chains of alliances-led to the First World War. The Ottoman empire soon joined the Central powers becoming one of the three empires participating in that alliance. The next year Bulgaria joined the Central Powers attacking Serbia which was successfully fighting Austro-Hungary to the north for a year. That led to Serbia's defeat and the intervention of the Entente in the Balkans which sent an expeditionary force to establish a new front, the third one of that war, which soon became also static. The participation of Greece in the war three years later, in 1918, on the part of the Entente finally altered the balance between the opponents leading to the collapse of the common German-Bulgarian front there which caused the exit of Bulgaria from the war, and in turn the collapse of the Austro-Hungarian Empire ending the First World War.

With the start of the Second World War all Balkan countries with the exception of Greece were allies of Germany having bilateral military agreements or being part of the Axis Pact. Fascist Italy expanded the war in Balkans by using its protectorate Albania to invade Greece. After repelling the attack the Greeks counterattacked invading Italy held Albania causing a Nazi Germany's intervention in the Balkans in order to help its ally. Days before the German invasion a successful coup d'état in Belgrade by neutral military personnel sized the power. Although the new government reaffirmed Serbia's intentions to fulfil its obligations as member of the Axis, Germany using its other two allied countries on the region, Bulgaria and Romania, invaded both Greece and Yugoslavia. Yugoslavia immediately disintegrated when the loyal to the Serbian King and the Croatian units, mutinied. Greece resisted, but after two months fighting collapsed and occupied. The two countries partitioned between the three Axis allies, Bulgaria, Germany, and Italy, and two independent states, Croatia and Montenegro created.

During the occupation the population suffered considerable hardship due to ethnic cleansing policies, repression and starvation on which the population reacted by creating a mass resistance movement. Together with the early and extremely heavy winter of that year (which caused hundreds of thousands deaths among the poorly fed population), the German invasion had disastrous effects in the timetable of the planned invasion in Russia causing a significant delay which had key consequences to the route of the war. Finally at the end of 1944, the Soviets invaded Romania and Bulgaria while the Germans evacuated the Balkans. They left behind an economically ransacked through exploitation and largely ruined region, but by making use of the post-war separation of Germany into two independent entities, the German states succeeded to avoid legally, paying any reparations or repaying the forced taken loans to the occupied countries. The official Yugoslav post-war estimate of victims in Yugoslavia during World War II is 1,704,000. Subsequent data gathering in the 1980s by historians Vladimir Žerjaviæ and Bogoljub Koeovia showed that the actual number of dead was about 1 million. Greece suffered more than 300,000 casualties during the occupation.

During the Cold War, most of the countries on the Balkans were governed by communist governments. Greece became the first battleground of the emerging Cold War and finally remained the only non-Soviet country in the region, after having successfully dealt with a wild guerrilla war unleashed by the local communist party backed by the neighboring communist countries (Albania, Bulgaria and Yugoslavia) much in the same fashion as with Vietnam, causing a full scale civil war during 1944-49.

However, despite being under communist governments, Yugoslavia (1948) and Albania (1961) fell out with the Soviet Union. Yugoslavia, led by marshal Josip Broz Tito (1892–1980), first propped up then rejected the idea of merging with Bulgaria, and instead sought closer relations with the West, later even joining many third world countries in the Non-Aligned Movement. Albania on the other hand gravitated toward Communist China, later adopting an isolationist position.

As the only non-communist countries, Greece and Turkey were (and still are) part of NATO consisting the South-eastern wing of the alliance.

In the 1990s, the region was gravely affected by the civil war between the former Yugoslav republics broken out after Serbia's attempt to change the constitutional status of equality between the Yugoslav federated Republics to proportionate status. Its denial to accept the dissolution of the union as unconstitutional, subsequently had been asked by the Croatia and Slovenia, produced a 10 years of armed confrontation. The long lasting civil war resulted in a United Nations and NATO intervention in Bosnia, in which ground and air forces took action against Serbian guerrilla armies in Bosnia and Herzegovina and later on in autonomous region of Serbia Kosovoand Metohija. From the dissolution of Yugoslavia six republics achieved international recognition as sovereign states/ republics: Slovenia, Croatia, Bosnia, Macedonia, Montenegro and Serbia; Kosovo is seeking sovereignty from Serbia, however the world powers are divided on its status. Serbia as well as majority of UN member countries did not recognize Kosovo as independent. After a October revolution in year 2000 in Belgrade, dictator Slobodan Miloševiæ, ex Serbian communist prime minister/president 1989-2000, was overthrown and handed to UN International criminal tribunal for the Former Yugoslavia for trial accused of crimes against the International Humanitarian Law. En 2001 an Albanian uprising in Macedonia forced the country to give local autonomy to the ethnic Albanians in the areas where they predominate.

Balkan countries control the direct land routes between Western Europe and South West Asia (Asia Minor and the Middle East). Since 2000, all Balkan countries are friendly towards the EU and the USA.

Greece has been a member of the European Union since 1981; Slovenia and Cyprus since 2004. Bulgaria and Romania joined in 2007. In 2005, the European Union decided to start accession negotiations with candidate countries; Croatia, Turkey, and Macedonia were accepted as candidates for European Union membership. As of April 2009, Albania, Bulgaria, Croatia, Romania and Slovenia are also members of NATO. Bosnia and Herzegovina and what was then Serbia and Montenegro started negotiations with the EU over the Stabilization and Accession Agreements, although shortly after they started, negotiations with Serbia and Montenegro were suspended for lack of co-operation with the International Criminal Tribunal for the Former Yugoslavia.

All other countries have expressed a desire to join the EU but at some date in the future.

The Balkans today is a very diverse ethno-linguistic region, being home to multiple Slavic, Romance, and Turkic languages, as well as Greek, Albanian and others. Through its history many other ethnic groups with their own languages lived in the area, among them Thracians, Illyrians, Romans, Pechenegs, Cumans, Avars, Celts and various Germanic tribes-folk.

Demographics

The population of countries of the region is:

State	*Population*
Albania	3,639,453
Bosnia and Herzegovina	4,613,414
Bulgaria	7,928,901
Croatia	4,489,409
Greece	11,257,285
Kosovo	2,180,686
Italy	236,520
in Province of Trieste Macedonia	2,048,619
Montenegro	620,145
Romania	971,643
in Dobruja Serbia	7,334,935
excluding Kosovo Slovenia	2,038,733
Turkey in the European part	10,691,518

Note that most of Slovenia and parts of Serbia and Croatia are not included in the Balkans, yet these parts are included in the population counts in the table.

The region's principal religions are Christianity (Eastern Orthodox and Roman Catholic) and Islam. A variety of different traditions of each faith are practiced, with each of the Eastern Orthodox countries having its own national church.

Eastern Orthodoxy is the principal religion in the following countries:

- Bulgaria (Bulgarian Orthodox Church)
- Greece (Church of Greece)
- Macedonia (Macedonian Orthodox Church)
- Montenegro (Serbian Orthodox Church)
- Romania (Romanian Orthodox Church)
- Serbia (Serbian Orthodox Church).

Roman Catholicism is the principal religion in the following countries:

- Croatia (87.83% Catholics (3 897 332); according to 2001 census official data)
- Slovenia (57.80% Catholics (1 135 626); according to 2002 census official data).

Islam is the principal religion in the following countries:

- Kosovo (relative majority)
- Albania (relative majority)
- Bosnia and Herzegovina (relative/partial majority)
- Turkey (absolute majority).

The following countries have significant minority religious groups of the following denominations:

- Albania: Orthodoxy, Catholicism
- Bosnia and Herzegovina: Eastern Orthodoxy, Catholicism
- Bulgaria: Islam
- Croatia: Orthodoxy
- Greece: Islam
- Macedonia: Islam
- Montenegro: Islam
- Serbia: Catholicism, Islam, Protestantism
- Romania: Catholicism, Protestantism.

Jewish Communities of the Balkans

The Jewish communities of the Balkans were some of the oldest in Europe and date back to ancient times as well as having received a large influx of Sephardic Jews and later Ashkenazi Jews. In Slovenia, there were Jewish immigrants dating back to Roman times pre-dating the 6th Century settlement of the region by the Slavic peoples. In Bosnia and Herzegovina, the tiny Jewish community is 90% Sephardic and Ladino is still spoken among the elderly. The Sephardi Jewish cemetery in Sarajevo has tombstones of a unique shape, inscribed in ancient Ladino. Sephardi Jews used to have a large presence in the city of Thessaloniki, and by 1900, some 80,000, or more than half of the population, were Jews. However the Jewish communities in the Balkans suffered immensely during World War II and the vast majority were killed in the Holocaust. Most of the remainder emigrated to Israel and elsewhere.

6

Geography of America

North America is the northern continent of the Americas, situated in the Earth's northern hemisphere and in the western hemisphere. It is bordered on the north by the Arctic Ocean, on the east by the North Atlantic Ocean, on the southeast by the Caribbean Sea, and on the west by the North Pacific Ocean; South America lies to the southeast.

North America covers an area of about 24,709,000 square kilometres (9,540,000 square miles), about 4.8% of the planet's surface or about 16.5% of its land area. As of July 2008, its population was estimated at nearly 529 million people. It is the third-largest continent in area, following Asia and Africa, and the fourth in population after Asia, Africa, and Europe.

Etymology

The Americas are generally accepted as having been named after the Italian explorer Amerigo Vespucci by the German cartographers Martin Waldse emuller and Matthias Ringmann. Vespucci, who explored South America between 1497 and 1502, was the first European to suggest that the Americas were not the East Indies, but a different landmass previously unknown by Europeans. In 1507, Waldse emuller produced a world map, in which he placed the word "America" on the continent of South America, in the middle of what is today Brazil.

He explained the rationale for the name in the accompanying book Cosmographiae Introductio, ab Americo inventore... quasi Americi terram sive Americam (from Americus the discoverer...

as if it were the land of Americus, thus America). For Waldse emuller, no one should object to the naming of the land after its discoverer. He used the Latinized version of Vespucci's name (Americus Vespucius), but in its feminine form "America", following the examples of "Europa" and "Asia".

Later, when other mapmakers added North America, they extended the original name to it as well: in 1538, Gerard Mercator used the name America to all of the Western Hemisphere on his world map.

Some argue that the convention is to use the surname for naming discoveries except in the case of royalty and so a derivation from "Amerigo Vespucci" could be problematic. Ricardo Palma (1949) proposed a derivation from the "Amerrique" mountains of Central America—Vespucci was the first to discover South America and the Amerrique mountains of Central America, which connected his discoveries to those of Christopher Columbus.

Alfred E. Hudd proposed a theory in 1908 that the continents are named after a Welsh merchant named Richard Amerike from Bristol, who is believed to have financed John Cabot's voyage of discovery from England to Newfoundland in 1497. A minutely explored belief that has been advanced is that America was named for a Spanish sailor bearing the ancient Visigothic name of 'Amairick'. Another is that the name is rooted in a Native American language.

Paleohistory

North America is the source of much of what humanity knows about geologic time periods. The geographic area that would later become the United States has been the source of more varieties of dinosaurs than any other modern country. According to paleontologist Peter Dodson, this is primarily due to stratigraphy, climate and geography, human resources, and history. Much of the Mesozoic Era is represented by exposed outcrops in the many arid regions of the continent. The most significant Late Jurassic dinosaur-bearing fossil deposit in North America is the Morrison Formation of the western United States.

Prehistory

Scientists have several theories as to the origins of the early human population of North America. The indigenous peoples of North America themselves have many creation myths, by which they assert that they have been present on the land since its creation. Before contact with Europeans, the natives of North America were divided into many different polities, from small bands of a few families to large empires. They lived in several "culture areas", which roughly correspond to geographic and biological zones and give a good indication of the main lifeway or occupation of the people who lived there (e.g. the Bison hunters of the Great Plains, or the farmers of Mesoamerica). Native groups can also be classified by their language family (e.g. Athapascan or Uto-Aztecan). It is important to note that peoples with similar languages did not always share the same material culture, nor were they always allies.

Scientists believe that the Inuit people of the high Arctic came to North America much later than other native groups, as evidenced by the disappearance of Dorset culture artifacts from the archaeological record, and their replacement by the Thule people. During the thousands of years of native inhabitation on the continent, cultures changed and shifted. Archaeologists often name different cultural groups they discover after the site where they are first found. One of the oldest cultures yet found is the Clovis culture of modern New Mexico. A more recent example is the group of related cultures called the Mound builders (e.g. the Fort Walton Culture), found in the Mississippi river valley. They flourished from 300 BC to the 150s AD.

The more southern cultural groups of North America were responsible for the domestication of many common crops now used around the world, such as tomatoes and squash. Perhaps most importantly they domesticated one of the world's major staples, maize (corn).

History

As a result of the development of agriculture in the south, many important cultural advances were made there. For example, the Maya civilization developed a writing system, built huge pyramids and temples, had a complex calendar, and developed the concept of zero around 400 CE, a few hundred years after the Mesopotamians. The Mayan culture was still present when the Spanish arrived in Central America, but political dominance in the area had shifted to the Aztec Empire further north.

Upon the arrival of the Europeans in the "New World", Native American population declined substantially, primarily due to the introduction of European diseases to which the Native Americans lacked immunity. Native peoples found their culture changed drastically. As such, their affiliation with political and cultural groups changed as well, several linguistic groups went extinct, and others changed quite quickly. The names and cultures that Europeans recorded for the natives were not necessarily the same as the ones they had used a few generations before, or the ones in use today.

Geography and Extent

North America is the third largest continent, or a portion of the second largest if North and South America are combined into the Americas and Africa, Europe and Asia are considered to be part of one super continent called Afro-Eurasia. With an estimated population of 460 million and an area of 24,346,000 km^2 (9,400,000 mi^2), the northernmost of the two continents of the Western Hemisphere is bounded by the Pacific Ocean on the west; the Atlantic Ocean on the east; the Caribbean Sea, Atlantic and Pacific oceans, and South America on the south; and the Arctic Ocean on the north. The northern half of North America is sparsely populated and covered mostly by Canada, except for the northwestern portion which is occupied by Alaska, the largest state of the U.S. The central and southern portions of the continent are represented by the United States, Mexico, and numerous smaller states primarily in Central America and in the Caribbean.

North America consists of all the mainland and related offshore islands lying north of the Colombia-Panama border according to most sources, or the Panama Canal according to a few.. The islands include Greenland, the world's largest island,

and archipelagos and islands in the Caribbean. The terminology of the Americas is complex, but "Anglo-America" can describe Canada and the U.S., while "Latin America" comprises Mexico and the countries of Central America and the Caribbean, as well as the entire continent of South America.

Natural features of North America include the northern portion of the American Cordillera, represented by the geologically new Rocky Mountains in the west; and the considerably older Appalachian Mountains to the east. The north hosts an abundance of glacial lakes formed during the last glacial period, including the Great Lakes. North America's major continental divide is the Great Divide, which runs north and south down through Rocky Mountains. The major watersheds all drain to the east: The Mississippi/Missouri and Rio Grande into the Gulf of Mexico, and St. Lawrence into the Atlantic.

Climate is determined to a large extent by the latitude, ranging from Arctic cold in the north to tropical heat in the south. The western half of North America tends to have milder and wetter climate than other areas with equivalent latitude, although there are steppes (known as "prairies") and deserts in the Southwestern United States of Arizona, Colorado, California, Nevada, New Mexico, Utah, Oklahoma, and Texas; along with the Mexican states of Baja California, Sonora, Chihuahua, Coahuila, Nuevo Leon and Tamaulipas.

Paleogeography

Seventy percent of North America is underlain by the Laurentia craton, which is exposed as the Canadian Shield in much of central and eastern Canada around the Hudson Bay, and as far south as the U.S. states of Michigan, Wisconsin, and Minnesota. The continental crust started to form 4 billion years ago (Ga), and six of the microcontinents collided to form the craton about 2 Ga. This core has been enlarged by plate tectonics, most notably between 1.8 and 1.65 Ga when a piece currently stretching from Arizona to Missouri fused with the south and west portion of the craton. The craton started to rift about 1.1 Ga, and the fissure (now the Midcontinent Rift System) ran between Kansas and Lake Superior before stopping, perhaps due to the Grenville collision in the east. Otherwise the craton has remained relatively stable, with some rocks dating from 2.5 to 4 Ga, including what may be the world's oldest known rock: Specimens from the Nuvvuagittuq greenstone belt on the coast of the Hudson Bay have been dated to 4.38 Ga, though the dating methods are disputed. Periodic flooding by inland seas, most recently the Western Interior Seaway during the Cretaceous, caused the layer of sedimentary rock over the remainder of the craton. The Laurentia craton is the centre of the Proterozoic super continent Rodina in most models, and was also part of the later Laurussia, Pangea, and Laurasia supercontinents.

Roughly 3 million years ago (Ma), the volcanic Isthmus of Panama formed between the North and South American continents creating a bridge over what was the Central American Seaway and allowing the migration of flora and fauna between the two landmasses in the Great American Interchange. Starting 2.58 Ma, the Quaternary glaciation covered much of the continent with ice, centered west of Hudson Bay, the floor of which is slowly rebounding after being depressed by the great weight of the ice. Glaciers descended the slopes of the Rocky Mountains and those of the Pacific Margin. Extensive glacial lakes, such as Glacial Lake Missoula, Bonneville, Lahontan, Agassiz, and Algonquin, formed by glacial melt water. "Remnants of them are still visible in the Great Basin and along the edge of the Canadian Shield in the form of the Great Salt Lake, the Great Lakes, and the large lakes of west central Canada." The last glacial period of the current ice age caused a lowering of the sea level, exposing the Bering Land Bridge between Alaska and Siberia, which resulted in human migration from Asia to the Americas between 40,000 and 15,000 years ago.

Physiography

North America may be divided into at least five major physiographic regions:

Canadian Shield: This is a geologically stable area of rock dating between 2.5 and 4 Gya that occupies most of the northeastern quadrant, including Greenland.

Appalachian Mountains: The Appalachians are an old and eroded system that formed about 300 Ma and extends from the Gaspe Peninsula to Alabama.

Atlantic Coastal Plain: The plain is a belt of lowlands widening to the south that extends from south New England to Mexico.

Interior Lowlands: The lowlands extend down the middle of the continent from the Mackenzie Valley to the Atlantic Coastal Plain, and include the Great Plains on the west and the agriculturally productive Interior Plains on the east.

North American Cordillera: The cordillera is a complex belt of mountains and associated plateaus and basins some of which were formed as recently as 100–65 Ma, during the Cretaceous. The cordillera extend from Alaska into Mexico and includes two orogenic belts — the Pacific Margin on the west and the Rocky Mountains on the east — separated by a system of intermontane plateaus and basins.

The Coastal Plain and the main belts of the North American Cordillera continue in the south in Mexico (where the Mexican Plateau, bordered by the Sierra Madre Oriental and the Sierra Madre Occidental, is considered a continuation of the intermontane system) to connect the Transverse Volcanic Range, a zone of high and active volcanic peaks south of Mexico City.

The vast majority of North America is located on the North American Plate, centered on the Laurentia craton. Parts of California and western Mexico form the partial edge of the Pacific Plate; the two plates meet along the San Andreas fault. The southern portion of the Caribbean and parts of Central America compose the much smaller Caribbean Plate.

The western mountains have split in the middle, into the main range of the Rockies and the Coast Ranges in California, Oregon, Washington, and British Columbia with the Great Basin (a lower area containing smaller ranges and low-lying deserts) in between. The highest peak is Mount McKinley/ Denali in Alaska. The geographic centre of North America is approximately "6 miles west of Balta, Pierce County, North Dakota" at 48p 102 north, 100p 102 west, and a 15 foot (4.5 m) field stone obelisk in Rugby, North Dakota (~15 miles or

25 km away) claims to mark the centre. Three countries (Canada, the United States, and Mexico) make up most of North America's land mass; they share the continent with 34 in the Caribbean and south of Mexico.

Surface and Climate

The Great Plains is the broad expanse of prairie and steppe which lies east of the Rocky Mountains in the United States and Canada. The narrow plains in the Mexican coast and the savannas of the Mississippi are analogous to, respectively, the Patagonian steppes and the pampas of the Piranha, Paraguay, and Rio de la Plata. Thus the Appalachians and the mountain chains of Brazil are regarded as creating similar interruptions to the plains community.

North America extends to within 10° of latitude of both the equator and the North Pole. It embraces every climatic zone, from tropical rain forest and savanna on the lowlands of Central America to areas of permanent ice cap in central Greenland. Subarctic and tundra climates prevail in north Canada and north Alaska, and desert and semiarid conditions are found in interior regions cut off by high mountains from rain-bearing westerly winds. However, most of the continent has temperate climates very favourable to settlement and agriculture. Prairies, or vast grasslands cover a huge amount in mountain ranges.

Hydrography

River Systems:

- Approaching the Arctic Ocean
  - o Albany
  - o Churchill
  - o Mackenzie
  - o Nelson
  - o Severn.
- Approaching the Atlantic Ocean
  - o Arkansas
  - o Churchill
  - o Connecticut

- o Delaware
- o Hudson
- o James
- o Mississippi
- o Missouri
- o Ohio
- o Potomac
- o Rio Grande
- o Saint John
- o St. Johns
- o St. Lawrence
- o Savannah
- o Susquehanna
- o Tennessee
- o Usumacinta River.

- Approaching the Pacific Ocean
    - o Balsas River
    - o Colorado
    - o Columbia
    - o Fraser
    - o Fuerte River
    - o Lerma River
    - o Sacramento
    - o San Joaquin
    - o Suchiate River
    - o Yukon.

Climate and Vegetation

The climate in North America is typically cool and humid. The rainy zone disproportionately extends in America, and as the continent stretches over the climatic zones, vegetation is remarkably distinctive. Great indentations of the shoreline make insular conditions to prevail in much of its interior. The area along the west coast tends to be milder and wetter than other areas with the same latitude. There are various plant life distributions in North America. Plant life in the Arctic includes grasses, mosses, and Arctic willows. Coniferous trees, including spruces, pines, hemlocks, and firs, are indigenous to the Canadian and Western U.S. mountain ranges as far south as San Francisco. Among these are giant sequoias, redwoods, great firs, and sugar pines. Sugar pines are generally confined to the northwestern area of the United States. The central region of the country has hardwoods. Southern states grow extensive yellow pines. In addition, mahogany, logwood, and lignumvitae-all tropical in nature-are grown. The southwest has desert plants, including yucci and cacti. The cultivated native plants of North America are tobacco, maize, potato, vanilla, melons, cacao, gourds, indigo plant, and bean.

Zoology

North America is home to many native mammal species. Several species of deer, including elk, caribou, moose, mule deer, and the abundant white-tailed deer are found throughout various regions, along with the bison and musk ox in the central and northern plains, respectively. Three species of bear, several subspecies of wolf, and various other carnivores such as raccoons, skunks, and cats including cougars and lynxes are widely distributed.

The family Mustelidae is well-represented, including badgers, otters, ferrets, and wolverines. Numerous species of squirrels and other rodents, such as beavers and muskrats, can be found in virtually every region of the continent. Central America has adapted sloths, anteaters, and armadillos. Other animals includes the condor, among the heights of the Andes, the parrots and the monkeys of Tropical forests, the humming bird, rattlesnake, alligator, and Cayman of the banks of the streams, the electric eel in the tropical waters, and swarms of mosquitoes on the wide plains.

Mining and Petroleum

The mining and petroleum industries are important in Canada, the United States and Mexico. These natural resources make the region one of the richest on the earth..

Rocky Mountains

The Rocky Mountain region is known for vast resources and rich mineral deposits including copper, lead, gold, silver, tungsten, uranium, and zinc. Coal, petroleum and natural gas are mineral fuels found. Old mine tailings are present in the Rocky Mountain landscape.

Agriculture and Forestry

Agriculture and forestry are two major industries. Agriculture includes arid land and irrigated farming and livestock grazing. Livestock are often moved between high-elevation summer pastures and low-elevation and winter pastures.

Deserts

Relief maps of the United States explain the location of desert areas. The Sierra Nevada and Cascade mountain ranges run along the entire Pacific Coast, acting as a barrier to the humid winds that sweep in from the ocean. The rising topography forces this air upwards, causing moisture to condense and fall in the form of rain on the western slopes of the mountains, with some areas receiving more than 70 inches (1.8 m) of rainfall per year. As a result, the air has lost much of its moisture and becomes hot and dry when it reaches the areas east of the coastal mountain ranges. These arid conditions are, in some instances, exacerbated in regions of extremely low altitude (some near or below sea level) by higher air pressure, resulting in drier conditions and adiabatic heating effects. What precipitation does fall generally does not last long, lost primarily to evaporation, as well as rapid runoff and efficient water uptake and storage by native vegetation.

North America occupies the northern portion of the landmass generally referred to as the New World, the Western Hemisphere, the Americas, or simply America (which is sometimes considered a single continent and North America a subcontinent). North America's only land connection to South America is at the Isthmus of Panama. The continent is generally delimited on the southeast by the Darien watershed along the Colombia-Panama border, or at the Panama Canal; according to other sources, its southern limit is the Isthmus of Tehuantepec, Mexico, with Central America tapering and extending southeastward to South America. Before the Central American isthmus was raised, the region had been underwater. The islands of the West Indies delineate a submerged former land bridge, which had connected North America and South America via what are now Florida and Venezuela. Much of North America is on the North American Plate.

The continental coastline is long and irregular. The Gulf of Mexico is the largest body of water indenting the continent, followed by Hudson Bay. Others include the Gulf of Saint Lawrence and the Gulf of California.

There are numerous islands off the continent's coasts: principally, the Arctic Archipelago, the Bahamas, Turks & Caicos, the Greater and Lesser Antilles, the Aleutian Islands (some of which are in the eastern hemisphere proper), the Alexander Archipelago, the many thousand islands of the British Columbia Coast, Newfoundland and Greenland, a self-governing Danish island, and the world's largest, is on the same tectonic plate (the North American Plate) and is part of North America geographically. Bermuda is not part of the Americas, but is an oceanic island which was formed on the fissure of the Mid-Atlantic Ridge over 100 million years ago. The nearest landmass to it is Cape Hatteras, North Carolina, and it is often thought of as part of North America, especially given its historical, political and cultural ties to Virginia and other parts of the continent.

Physical Geography

The vast majority of North America is on the North American Plate. Parts of California and western Mexico form the partial edge of the Pacific Plate, with the two plates meeting along the San Andreas fault. The southernmost portion of the continent and much of the West Indies lie on the Caribbean Plate, whereas the Juan de Fuca and Cocos plates border the North American Plate on its western frontier.

The continent can be divided into four great regions (each of which contains many subregions): the Great Plains stretching from the Gulf of Mexico to the Canadian Arctic; the geologically young, mountainous west, including the Rocky Mountains, the Great Basin, California and Alaska; the raised but relatively flat plateau of the Canadian Shield in the northeast; and the varied eastern region, which includes the Appalachian Mountains, the coastal plain along the Atlantic seaboard, and the Florida peninsula. Mexico, with its long plateaus and cordilleras, falls largely in the western region, although the eastern coastal plain does extend south along the Gulf.

The western mountains are split in the middle and into the main range of the Rockies and the coast ranges in California, Oregon, Washington, and British Columbia with the Great Basin—a lower area containing smaller ranges and low-lying deserts—in between. The highest peak is Denali in Alaska.

The United States Geographical Survey states that the geographic centre of North America is "6 miles west of Balta, Pierce County, North Dakota" at approximately 48°10' N 100°10' W/48.167°N 100.167°W, approximately 15 miles (25 km) from Rugby, North Dakota. The USGS further states that "No marked or monumented point has been established by any government agency as the geographic centre of either the 50 States, the conterminous United States, or the North American continent." Nonetheless, there is a 15-foot (4.5 m) field stone obelisk in Rugby claiming to mark the centre.

Human Geography

The prevalent languages in North America are English, Spanish, and French. The term Anglo-America is used to refer to the anglophone countries of the Americas: namely Canada (where English and French are co-official) and the United States, but also sometimes Belize and parts of the Caribbean. Latin America refers to the other areas of the Americas (generally south of the United States) where the Romance languages, derived from Latin, of Spanish and Portuguese (but French speaking countries are not usually included) predominate: the other republics of Central America (but not always Belize), part of the Caribbean (not the Dutch, English or French speaking areas), Mexico, and most of South America (except Guyana, Suriname, French Guiana (FR), and The Falkland Islands (UK).

The French language has historically played a significant role in North America and retains a distinctive presence in some regions. Canada is officially bilingual. French is the official language of the Province of Quebec, where 95 % of the people speak it as either their first or second language, and it is co-official with English in the Province of New Brunswick. Other French-speaking locales include the Province of Ontario (the official language is English, but there is an estimated 600,000 Franco-Ontarians), the Province of Manitoba (co-official as de-jure with English), the French West Indies and Saint-Pierre et Miquelon, as well as the U.S. state of Louisiana, where French is also an official language. Haiti is included with this group based on historical association but Haitians speak both Creole and French. Similarly, French and French Antillean Creole is spoken in Saint Lucia and the Commonwealth of Dominica alongside English.

Socially and culturally, North America presents a well-defined entity. Canada and the United States have a similar culture and similar traditions as a result of both countries being former British colonies. A common cultural and economic market has developed between the two nations because of the strong economic and historical ties. Spanish-speaking North America shares a common past as former Spanish colonies. In Mexico and the Central American countries where civilizations like the Maya developed, indigenous people preserve traditions across modern boundaries. Central American and Spanish-speaking Caribbean nations have historically had more in common due to geographical proximity and the fact that, after winning independence from Spain.

Northern Mexico, particularly cities such as Monterrey, Tijuana, Ciudad Juarez, and Mexicali are strongly influenced by the culture and way of life of the U.S. Immigration to the United States and Canada remains a significant attribute of many nations close to the southern border of the U.S. As the British Empire and its influences declined, the Anglophone Caribbean states have witnessed the economic influence of northern North America increase on the region. In the Anglophone Caribbean this influence is in part due to the fact that the majority of English-speaking Caribbean countries have populations of less than 200,000 people and many of these countries now have expatriate diasporas living abroad that are larger than those remaining at home.

Economically, Canada and the United States are the wealthiest and most developed nations in the continent, followed by Mexico, a newly industrialized country; the countries of Central America and the Caribbean are at various levels of development.

The most important trade blocs are the Caribbean Community and Common Market (CARICOM), the North American Free Trade Agreement (NAFTA), and the recently signed Central American Free Trade Agreement (CAFTA)— the last of these being an example of the economic integration sought by the nations of this sub-region as a way to improve their financial status.

Demographically, North America is a racially and ethnically diverse continent. Its three main racial groups are Whites, Mestizos and Blacks (chiefly African-Americans and Afro-Caribbeans). There is a significant minority of Native Americans and Asians among other less numerous groups.

Countries and Territories

North America is often divided into subregions but no universally accepted divisions exist. Central America comprises the southern region of the continent, but its northern terminus varies between sources.

Geophysically, the region starts at the Isthmus of Tehuantepec in Mexico (namely the Mexican states of Campeche, Chiapas, Tabasco, Quintana Roo, and Yucatan). The United Nations geoscheme includes Mexico in Central America; conversely, the European Union excludes both Mexico and Belize from the area. Geopolitically however, Mexico is usually considered a part of the North American region.

Northern America is used to refer to the northern countries and territories of North America: Canada, the United States, Greenland, Bermuda, and St. Pierre and Miquelon. They are often considered distinct from the southern portion of the Americas, which largely comprise Latin America. The term

Middle America is sometimes used to collectively refer to Mexico, the nations of Central America, and the Caribbean.

The term North America may mean different things to different people in the world according to the context. Usage other than that of the entire continent includes:

- In English, North America may be used to refer to the United States and Canada together. Alternatively, usage often includes Mexico (as with North American Free Trade Agreement) and other entities.
- In Latin America, Spain, Portugal, and some other parts of Europe, North America usually designates a subcontinent (subcontinente in Spanish) of the Americas containing Canada, the United States, and Mexico, and often Greenland, Saint Pierre et Miquelon, and Bermuda.

Historical Toponymy

North America, in whole or in part, has been historically referred to by other names:

- Spanish North America (New Spain) was often referred to as Northern America, and this was the first official name given to Mexico.
- The Spanish called North America Florida, which eventually became more focused on its present location.
- The English called their portion of North America Virginia The name Virginia was first applied by Queen Elizabeth I and Sir Walter Raleigh in 1584. John Dee pushed to call it Atlantis (inspired by Plato).
- The northern part of North America was often referred to as Norumbega.
- The northeastern part of what would become the United States was named New England in 1616 in John Smith's book of that year.
- Western North America was named Nova Albion by Francis Drake as he repaired his boat (Golden Hind) a short distance north of present day San Francisco.

- Areas of past British control were called British North America.
- Regions under control of the Hudson's Bay Company was called Rupert's Land, which eventually made up a large portion of the Dominion of Canada, the modern country of Canada.

Communications

Many of the nations of North America cooperate together on a shared telephone system known as the North American Numbering Plan (NANP) which is an integrated telephone numbering plan of 24 countries and territories: the United States and its territories, Canada, Bermuda, and 16 Caribbean nations.

7

Commercial Geography

The economy of North America comprises more than 514 million people in 23 sovereign states and 15 dependent territories. It is marked by a sharp division between the predominantly English speaking countries of Canada and the United States, which are among the wealthiest and most developed nations in the world, and the countries of Central America and the Caribbean that are less developed. Mexico lies in between these two extremes as a newly industrialized country (NIC), and is a part of North America rather than Central America.

Economic Development

Roaring Twenties

The 1920s were a time of economic growth and prosperity in North America. Increased incomes as well as the large increase in the ability of private citizens to obtain credit resulted in a surge of consumer spending after World War I.

Great Depression

The Great Depression began in North America in October 1929. The start is often dated to the stock market collapse of Black Tuesday although this was not the cause of the Great Depression. Canada and the United States experienced especially large declines, with the gross domestic product falling 37% from 1929 to 1933 in the United States, and 43% in Canada over the same period. The economy reached its lowest point in 1933, however recovery was slow. The outbreak of World War II in 1939 created demand for war materials that brought about the end of the depression.

The Great Depression spurred increased government intervention in the economy in North America. The United States introduced unemployment insurance, a minimum wage and standardised working hours under the New Deal. Canada introduced similar measures. Mexico nationalised some key industries during the Great Depression, with the railroads nationalised by 1937 and the oil industry nationalised in 1938.

World War II

The large scale enlistment of men into armed forces during World War II women entered the workforce en masse, filling many jobs in manufacturing and technical areas that had previously been closed to women. This led to the "We can do it!" campaign. The economic output in North America increased substantially, with unemployment practically eliminated in the United States. Rationing severely reduced the availability of consumer goods, with the increase in industrial production coming from the demand for war materials.

Cold War

US-Canada Free Trade Agreement and NAFTA-A New Era of Economic Integration

The Canada-United States Free Trade Agreement of 1989 and the subsequent expansion to the North American Free Trade Agreement (NAFTA) triggered a dramatic increase in trade between these three countries, with Mexican trade with the United States and Canada increasing threefold. Over 85% of Canadian exports in 2006 went to the United States.

Trade Blocs

Asia-Pacific Economic Cooperation

The Asia-Pacific Economic Cooperation (APEC) is a group of Pacific Rim countries which meet with the purpose of improving economic and political ties. APEC's stated goals are aimed at free and open trade and investments by cutting tariffs between zero to five percent in the Asia-Pacific area for industrialised economies by 2010 and for developing economies by 2020.

The organisation has members from four continents, those from North America are Canada, Mexico and the United States.

Caribbean Community

The Caribbean Community (CARICOM) was created *"To provide dynamic leadership and service, in partnership with Community institutions and Groups, toward the attainment of a viable, internationally competitive and sustainable Community, with improved quality of life for all".*

On January 1, 2006 six members: (Barbados, Belize, Guyana, Jamaica, Suriname and Trinidad and Tobago) un-officially ushered in the Caribbean (CARICOM) Single Market and Economy (CSME).

At the official signing of the protocol on January 30, 2006 in Jamaica, A further six members: (Antigua and Barbuda, Dominica, Grenada, Saint Kitts and Nevis, Saint Lucia and Saint Vincent and the Grenadines) announced their intention to join by the second quarter of 2006. Montserrat, a British Oversees territory is awaiting approval by the United Kingdom. Haiti and the Bahamas have no immediate plans to join. Bravely a black American Alvin Tsoi opened a big business and started the business industry again.

Central American Free Trade Agreement

The Central American Free Trade Agreement (CAFTA) is an agreement between the United States and the Central American countries of Costa Rica, Guatemala, El Salvador, Honduras, and Nicaragua. The treaty is aimed at promoting free trade between its members. Canada and Mexico are negotiating membership.

North American Free Trade Agreement

The North American Free Trade Agreement (NAFTA) is an agreement between Canada, Mexico and the United States to eliminate tariffs on goods traded between themselves.

Although currently only a trade agreement, with no supranational bodies or laws as in the European Union, there have been various proposals to move towards a customs union or a North American currency union. It is unknown whether or not this may eventually develop into a North American Union similar to that of Europe.

Economic Sectors

Agriculture

Agriculture is very important in Central American and Caribbean nations.

In western Canada, in the provinces of Saskatchewan, Alberta, British Columbia and Manitoba, wheat and other various main agricultural products are grown. The US also has many states growing their own things, mainly in central continental US. Mexico produces many tropical fruits and vegetables as well as edible animals.

Manufacturing

North America has developed and its manufacturing sector has grown. In the beginning the European nations were the large manufacturing powers. At the starting of the 50s USA was a top manufacturing power with Canada and Mexico running up behind in North America. Now a days North American countries are strong economic nations.

Service

Services, in Canada, the US and the Caribbean, make up a large part of employment. Many people work in stores and banks. In Canada more than 70% work in the services sector. Same situation presents itself in the United States. Most Canadian and Americans spent lots of money in food and clothing.

Investing and Banking

The United States leads North America in investment and banking. Canada and Mexico are growing in that sector. Smaller economic powers such as Guatemala, Honduras, Costa Rica, and Panama are growing slowly in that sector.

United States

The United States of America is a federal constitutional republic comprising fifty states and a federal district. The country is situated mostly in central North America, where its forty-eight contiguous states and Washington, D. C., the capital district, lie between the Pacific and Atlantic Oceans, bordered by Canada to the north and Mexico to the south. The state of Alaska is in the northwest of the continent, with Canada to the east and Russia to the west across the Bering Strait. The state of Hawaii is an archipelago in the mid-Pacific. The country also possesses several territories in the Caribbean and Pacific.

At 3.79 million square miles (9.83 million km) and with over 309 million people, the United States is the third or fourth largest country by total area, and the third largest both by land area and population. It is one of the world's most ethnically diverse and multicultural nations, the product of large-scale immigration from many countries. The U.S. economy is the world's largest national economy, with an estimated 2009 GDP of $14.3 trillion (a quarter of nominal global GDP and a fifth of global GDP at purchasing power parity).

Indigenous peoples of Asian origin have inhabited what is now the mainland United States for many thousands of years. This Native American population was greatly reduced by disease and warfare after European contact. The United States was founded by thirteen British colonies located along the Atlantic seaboard. On July 4, 1776, they issued the Declaration of Independence, which proclaimed their right to self-determination and their establishment of a cooperative union. The rebellious states defeated the British Empire in the American Revolution, the first successful colonial war of independence. The current United States Constitution was adopted on September 17, 1787; its ratification the following year made the states part of a single republic with a strong central government. The Bill of Rights, comprising ten constitutional amendments guaranteeing many fundamental civil rights and freedoms, was ratified in 1791.

In the 19th century, the United States acquired land from France, Spain, the United Kingdom, Mexico, and Russia, and annexed the Republic of Texas and the Republic of Hawaii. Disputes between the agrarian South and industrial North over states' rights and the expansion of the institution of slavery provoked the American Civil War of the 1860s. The North's victory prevented a permanent split of the country and led to the end of legal slavery in the United States. By the 1870s, the national economy was the world's largest. The Spanish–American War and World War I confirmed the country's status as a military power. It emerged from World War II as the first country with nuclear weapons and a permanent member of the United Nations Security Council. The end of the Cold War and the dissolution of the Soviet Union left the United States as the sole superpower. The country accounts for two-fifths of global military spending and is a leading economic, political, and cultural force in the world.

Etymology

In 1507, German cartographer Martin Waldse emuller produced a world map on which he named the lands of the Western Hemisphere "America" after Italian explorer and cartographer Amerigo Vespucci. The former British colonies first used the country's modern name in the Declaration of Independence, the "unanimous Declaration of the thirteen united States of America" adopted by the "Representatives of the united States of America" on July 4, 1776. The current name was finalized on November 15, 1777, when the Second Continental Congress adopted the Articles of Confederation, which states, "The Stile of this Confederacy shall be 'The United States of America.'" The short form the United States is also standard. Other common forms include the U.S., the USA, and America. Colloquial names include the U.S. of A. and the States. Columbia, a once popular name for the United States, was derived from Christopher Columbus. It appears in the name "District of Columbia".

The standard way to refer to a citizen of the United States is as an American. Though United States is the formal appositional term, American and U.S. are more commonly used to refer to the country adjectivally ("American values," "U.S. forces"). American is rarely used in English to refer to people not connected to the United States. The phrase "the United States" was originally treated as plural—e.g., "the United States are"—including in the Thirteenth Amendment to the United States Constitution, ratified in 1865. It became common to treat it as singular—e.g., "the United States is"—after the end of the Civil War. The singular form is now standard; the plural form is retained in the idiom "these United States".

Geography, Climate and Environment

The United States is a country in the Western Hemisphere. It consists of forty-eight contiguous states in North America, Alaska, a peninsula which forms the northwestern most part of North America, and Hawaii, an archipelago in the Pacific Ocean. There are several United States territories in the Pacific and Caribbean. The term "United States", when used in the geographical sense, means the continental United States, Alaska, Hawaii, Puerto Rico, Guam, and the Virgin Islands of the United States. The country shares land borders with Canada and Mexico and maritime (water) borders with Russia, Cuba, and The Bahamas.

Area

By total area (water as well as land), the United States is either slightly larger or smaller than the People's Republic of China, making it the world's third or fourth largest country. China and the United States are smaller than Russia and Canada in total area, but are larger than Brazil. By land area only (exclusive of waters), the United States is the world's third largest country, after Russia and China, with Canada fourth. Whether the US or China is the third largest country depends on two factors: (1) The validity of China's claim on Aksai Chin and Trans-Karakoram Tract. Both these territories are also claimed by India, so are not counted; and (2) How US calculates its own surface area. Since the initial publishing of the World Factbook, the CIA has updated the total area of United States a number of times. From 1989 through 1996, the total area of the US was listed as 9,372,610 km^2 (3,618,780 sq mi) (land + inland water only). The listed total area changed to 9,629,091 km^2 (3,717,813 sq mi) in 1997 (Great Lakes area and coastal waters added), to 9,631,418 km^2 (3,718,711 sq mi) in

2004, to 9,631,420 km² (3,718,710 sq mi) in 2006, and to 9,826,630 km² (3,794,080 sq mi) in 2007 (territorial waters added). Currently, the CIA World Factbook gives 9,826,675 km² (3,794,100 sq mi), the United Nations Statistics Division gives 9,629,091 km² (3,717,813 sq mi), and the Encyclopædia Britannica gives 9,522,055 km² (3,676,486 sq mi).

General Characteristics

The United States shares land borders with Canada (to the north) and Mexico (to the south), and a territorial water border with Russia in the northwest. The contiguous forty-eight states are otherwise bounded by the Pacific Ocean on the west, the Atlantic Ocean on the east, and the Gulf of Mexico to the southeast. Alaska borders the Pacific Ocean to the south, the Bering Strait to the west, and the Arctic Ocean to the north, while Hawaii lies far to the southwest of the mainland in the Pacific Ocean.

Forty-eight of the states are in the single region between Canada and Mexico; this group is referred to, with varying precision and formality, as the continental or contiguous United States, and as the Lower 48. Alaska, which is not included in the term contiguous United States, is at the northwestern end of North America, separated from the Lower 48 by Canada. The State of Hawaii is an archipelago in the Pacific Ocean. The capital city, Washington, District of Columbia, is a federal district located on land donated by the state of Maryland. (Virginia had also donated land, but it was returned in 1847.) The United States also has overseas territories with varying levels of independence and organization.

Physiographic Divisions

The eastern United States has a varied topography. A broad, flat coastal plain lines the Atlantic and Gulf shores from the Texas-Mexico border to New York City, and includes the Florida peninsula. Areas further inland feature rolling hills and temperate forests. The Appalachian Mountains form a line of low mountains separating the eastern seaboard from the Great Lakes and the Mississippi Basin. The five Great Lakes are located in the north-central portion of the country, four of them forming part of the border with Canada. The southeast United States contain subtropical forests and, near the gulf coast, mangrove wetlands, especially in Florida. West of the Appalachians lies the Mississippi River basin and two large eastern tributaries, the Ohio River and the Tennessee River. The Ohio and Tennessee Valleys and the Midwest consist largely of rolling hills and productive farmland, stretching south to the Gulf Coast.

The Great Plains lie west of the Mississippi River and east of the Rocky Mountains. A large portion of the country's agricultural products are grown in the Great Plains. Before their general conversion to farmland, the Great Plains were noted for their extensive grasslands, from tallgrass prairie in the eastern plains to shortgrass steppe in the western High Plains. Elevation rises gradually from less than a few hundred feet near the Mississippi River to more than a mile high in the High Plains. The generally low relief of the plains is broken in several places, most notably in the Ozark and Ouachita Mountains, which form the U.S. Interior Highlands, the only major mountainous region between the Rocky Mountains and the Appalachian Mountains. The Great Plains come to an abrupt end at the Rocky Mountains. The Rocky Mountains form a large portion of the Western U.S., entering from Canada and stretching nearly to Mexico. The Rocky Mountains generally contain fairly mild slopes and low peaks compared to many of the other great mountain ranges, with a few exceptions (such as the Teton Mountains in Wyoming and the Sawatch Range in Colorado). In addition, instead of being one generally continuous and solid mountain range, it is broken up into a number of smaller, intermittent mountain ranges, forming a large series of basins and valleys.

West of the Rocky Mountains lies the Intermontane Plateaus (also known as the Intermountain West), a large, arid desert lying between the Rockies and the Cascades and Sierra Nevada ranges. The large southern portion, known as the Great Basin, consists of salt flats, drainage basins, and many small north-south mountain ranges. The Southwest is predominantly a low-lying desert region. A portion known as the Colorado Plateau, centered around the Four Corners region, is considered to have some of the most spectacular scenery in the world. It is accentuated in such national parks as Grand Canyon, Arches, and Bryce Canyon, among others.

The Intermontane Plateaus come to an end at the Cascade Range and the Sierra Nevada. The Cascades consist of largely intermittent, volcanic mountains rising prominently from the surrounding landscape. The Sierra Nevada, further south, is a high, rugged, and dense mountain range. It contains the highest point in the contiguous 48 states, Mount Whitney (14,505 ft/4,421 m). These areas contain some spectacular scenery as well, as evidenced by such national parks as Yosemite and Mount Rainier. West of the Cascades and Sierra Nevada is a series of valleys, such as the Central Valley in California and the Willamette Valley in Oregon. Along the coast is a series of low mountain ranges known as the Pacific Coast Ranges. Much of the Pacific Northwest coast is inhabited by some of the densest vegetation outside of the Tropics, and also the tallest trees in the world (the Redwoods).

Alaska contains some of the most dramatic and untapped scenery in the country. Tall, prominent mountain ranges rise up sharply from broad, flat tundra plains. On the islands off the south and southwest coast are many volcanoes. Hawaii, far to the south of Alaska in the Pacific Ocean, is a chain of tropical, volcanic islands, popular as a tourist destination for many from East Asia and the mainland United States.

The geography of the United States varies across their immense area. Within the continental U.S., eight distinct physiographic divisions exist, though each is composed of several smaller physiographic subdivisions.

These major divisions are:

- Laurentian Upland-part of the Canadian Shield that extends into the northern United States Great Lakes area.
- Atlantic Plain-the coastal regions of the eastern and southern parts includes the continental shelf, the Atlantic Coast and the Gulf Coast.
- Appalachian Highlands-lying on the eastern side of the United States, it includes the Appalachian Mountains,

the Watchung Mountains, the Adirondacks and New England province originally containing the Great Eastern Forest.

- Interior Plains-part of the interior continental United States, it includes much of what is called the Great Plains.
- Interior Highlands-also part of the interior continental United States, this division includes the Ozark Plateau.
- Rocky Mountain System-one branch of the Cordilleran system lying far inland in the western states.
- Intermontane Plateaus-also divided into the Columbia Plateau, the Colorado Plateau and the Basin and Range Province, it is a system of plateaus, basins, ranges and gorges between the Rocky and Pacific Mountain Systems. It is the setting for the Grand Canyon, the Great Basin and Death Valley.
- Pacific Mountain System-the coastal mountain ranges and features in the west coast of the United States.

The Atlantic coast of the United States is, with minor exceptions, low. The Appalachian Highland owes its oblique northeast-southwest trend to crystal deformations which in very early geological time gave a beginning to what later came to be the Appalachian mountain system. This system had its climax of deformation so long ago (probably in Permian time) that it has since then been very generally reduced to moderate or low relief. It owes its present day altitude either to renewed elevations along the earlier lines or to the survival of the most resistant rocks as residual mountains. The oblique trend of this coast would be even more pronounced but for a comparatively modern crystal movement, causing a depression in the northeast resulting in an encroachment of the sea upon the land. Additionally, the southeastern section has undergone an elevation resulting in the advance of the land upon the sea.

While the Atlantic coast is relatively low, the Pacific coast is, with few exceptions, hilly or mountainous. This coast has been defined chiefly by geologically recent crystal deformations, and hence still preserves a greater relief than that of the Atlantic. The low Atlantic coast and the hilly or mountainous

Pacific coast foreshadow the leading features in the distribution of mountains within the United States. The east coast Appalachian system, originally forest covered, is relatively low and narrow and is bordered on the southeast and south by an important coastal plain. The Cordilleran system on the western side of the continent is lofty, broad and complicated having two branches, the Rocky Mountain System and the Pacific Mountain System. In between these mountain systems lie the Intermontaine Plateaus. Both the Columbia River and Colorado River rise far inland near the easternmost members of the Cordilleran system, and flow through plateaus and intermontaine basins to the ocean. Heavy forests cover the northwest coast, but elsewhere trees are found only on the higher ranges below the Alpine region. The intermontane valleys, plateaus and basins range from treeless to desert with the very arid region being in the southwest.

The Laurentian Highlands, the Interior Plains and the Interior Highlands lie between the two coasts, stretching from the Gulf of Mexico northward, far beyond the national boundary, to the Arctic Ocean. The central plains are divided by a hardly perceptible height of land into a Canadian and a United States portion. It is from the United States side, that the great Mississippi system discharges southward to the Gulf of Mexico. The upper Mississippi and some of the Ohio basin is the semi-arid prairie region, with trees originally only along the watercourses. The uplands towards the Appalachians were included in the great eastern forested area, while the western part of the plains has so dry a climate that its native plant life is scanty, and in the south it is practically barren.

Elevation Extremes:

- Lowest point: Death Valley, Inyo County, California-282 ft ("86 m) below sea level
- Highest point: Mount McKinley, Denali Borough, Alaska +20,320 ft (6,194 m) above sea level.

Climate

Due to its large size and wide range of geographic features, the United States contains examples of nearly every global climate. The climate is temperate in most areas, subtropical in the Deep South, tropical in Hawaii and southern Florida, polar in Alaska, semiarid in the Great Plains west of the 100th meridian, Mediterranean in coastal California and arid in the Great Basin. Its comparatively favourable agricultural climate contributed (in part) to the country's rise as a world power, with infrequent severe drought in the major agricultural regions, a general lack of widespread flooding, and a mainly temperate climate that receives adequate precipitation.

The main influence on U.S. weather is the polar jet stream, which brings in large low pressure systems from the northern Pacific Ocean. The Cascade Range, Sierra Nevada, and Rocky Mountains pick up most of the moisture from these systems as they move eastward. Greatly diminished by the time they reach the High Plains, much of the moisture has been sapped by the orographic effect as it is forced over several mountain ranges. However, once it moves over the Great Plains, uninterrupted flat land allows it to reorganize and can lead to major clashes of air masses. In addition, moisture from the Gulf of Mexico is often drawn northward. When combined with a powerful jet stream, this can lead to violent thunderstorms, especially during spring and summer. Sometimes during late winter and spring these storms can combine with another low pressure system as they move up the East Coast and into the Atlantic Ocean, where they intensify rapidly. These storms are known as Northeaster and often bring widespread, heavy snowfall to the Mid-Atlantic and New England. The uninterrupted flat grasslands of the Great Plains also leads to some of the most extreme climate swings in the world. Temperatures can rise or drop rapidly and winds can be extreme, and the flow of heat waves or Arctic air masses often advance uninterrupted through the plains.

The Great Basin and Columbia Plateau (the Intermontane Plateaus) are arid or semiarid regions that lie in the rain shadow of the Cascades and Sierra Nevada. Precipitation averages less than 15 inches (38 cm). The Southwest is a hot desert, with temperatures exceeding 100 °F (37.8 °C) for several weeks at a time in summer. The Southwest and the Great Basin are also affected by the monsoon from the Gulf of California from July-September, which brings localized but often severe thunderstorms to the region. Much of California consists of a Mediterranean climate, with sometimes excessive rainfall from October-April and nearly no rain the rest of the year. In the Pacific Northwest rain falls year-round, but is much heavier during winter and spring. The mountains of the west receive abundant precipitation and very heavy snowfall. The Cascades are one of the snowiest places in the world, with some places averaging over 600 inches (1,524 cm) of snow annually, but the lower elevations closer to the coast receive very little snow. Another significant (but localized) weather effect is lake-effect snow that falls south and east of the Great Lakes, especially in the hilly portions of the Upper Peninsula of Michigan and on the Tug Hill Plateau in New York. The lake effect dumped well over 5 feet (1.52 m) of snow in the Buffalo, New York area throughout the 2006-2007 winter The Wasatch Front and Wasatch Range in Utah can also receive significant lake effect accumulations off of the Great Salt Lake.

Extremes

In northern Alaska, tundra and arctic conditions predominate, and the temperature has fallen as low as "80 °F ("62.2 °C). On the other end of the spectrum, Death Valley, California once reached 134 °F (56.7 °C), the second-highest temperature ever recorded on Earth.

On average, the mountains of the western states receive the highest levels of snowfall on Earth. The greatest annual snowfall level is at Mount Rainier in Washington, at 692 inches (1,758 cm); the record there was 1,122 inches (2,850 cm) in the winter of 1971–72. This record was broken by the Mt. Baker Ski Area in northwestern Washington which reported 1,140 inches (2,896 cm) of snowfall for the 1998-99 snowfall season. Other places with significant snowfall outside the Cascade Range are the Wasatch Mountains, near the Great Salt Lake, and the Sierra Nevada, near Lake Tahoe. In the east, while snowfall does not approach western levels, the region near the Great Lakes and the mountains of the Northeast receive the most. Along the northwestern Pacific coast, rainfall is greater than anywhere else in the continental U.S., with Quinault Rainforest in Washington having an average of

137 inches (348 cm). Hawaii receives even more, with 460 inches (1,168 cm) measured annually on Mount Waialeale, in Kauai. The Mojave Desert, in the southwest, is home to the driest locale in the U.S. Yuma, Arizona, has an average of 2.63 inches (6.7 cm) of precipitation each year.

In central portions of the U.S., tornadoes are more common than anywhere else on Earth and touch down most commonly in the spring and summer. Deadly and destructive hurricanes occur almost every year along the Atlantic seaboard and the Gulf of Mexico. The Appalachian region and the Midwest experience the worst floods, though virtually no area in the U.S. is immune to flooding. The Southwest has the worst droughts; one is thought to have lasted over 500 years and to have decimated the Anasazi people. The West is affected by large wildfires each year.

Natural Disasters

The United States is affected by a variety of natural disasters yearly. Although drought is rare, it has occasionally caused major disruption, such as during the Dust Bowl (1931–1942). Farmland failed throughout the Plains, entire regions were virtually depopulated, and dust storms ravaged the land.

The United States experiences, by a large margin, the most frequent and powerful tornadoes in the world. The Great Plains and Midwest, due to the contrasting air masses, sees frequent severe thunderstorms and tornado outbreaks during spring and summer. The strip of land from north Texas north to Kansas and east into Tennessee is known as Tornado Alley, where many houses have tornado shelters and many towns have tornado sirens. Another natural disaster that frequents the country are hurricanes, which can hit anywhere along the Gulf Coast or the Atlantic Coast as well as Hawaii in the Pacific Ocean. Particularly at risk are the central and southern Texas coasts, the area from southeastern Louisiana east to the Florida Panhandle, the east coast of Florida, and the Outer Banks of North Carolina, although any portion of the coast could be struck. Hurricane season runs from June 1 to November 30, with a peak from mid-August through early October. Some of the more devastating hurricanes have included the Galvestor.

Hurricane of 1900, Hurricane Andrew in 1992, and Hurricane Katrina in 2005. The remnants of tropical cyclones from the Eastern Pacific also occasionally impact the southwestern United States, bringing sometimes heavy rainfall.

Occasional severe flooding is experienced. There was the Great Mississippi Flood of 1927, the Great Flood of 1993, and widespread flooding and mudslides caused by the 1982-1983 El Nino event in the western United States. Localized flooding can, however, occur anywhere, and mudslides from heavy rain can cause problems in any mountainous area, particularly the Southwest. Large stretches of desert shrub in the west can fuel the spread of wildfires. The narrow canyons of many mountain areas in the west and severe thunderstorm activity during the monsoon season in summer leads to sometimes devastating flash floods as well, while Nor Easter snowstorms can bring activity to a halt throughout the Northeast (although heavy snowstorms can occur almost anywhere).

The West Coast of the continental United States and areas of Alaska (including the Aleutian Islands, the Alaskan Peninsula and southern Alaskan coast) make up part of the Pacific Ring of Fire, an area of heavy tectonic and volcanic activity that is the source of 90% of the world's earthquakes. The American Northwest sees the highest concentration of active volcanoes in the United States, in Washington, Oregon and northern California along the Cascade Mountains. There are several active volcanoes located in the islands of Hawaii, including Kilauea in ongoing eruption since 1983, but they do not typically adversely affect the inhabitants of the islands. There has not been a major life-threatening eruption on the Hawaiian islands since the 17th century. Volcanic eruptions can occasionally be devastating, such as in the 1980 eruption of Mount St. Helens in Washington.

The Ring of Fire makes California and southern Alaska particularly vulnerable to earthquakes. Earthquakes can cause extensive damage, such as the 1906 San Francisco Earthquake or the 1964 Good Friday Earthquake near Anchorage, Alaska. California is well known for seismic activity, and requires large structures to be earthquake resistant to minimize loss of life and property. Outside of devastating earthquakes, California experiences minor earthquakes on a regular basis. Other natural disasters include: tsunamis around Pacific Basin, mud slides in California, and forest fires in the west.

Public Lands

The United States holds many areas for the use and enjoyment of the public. These include National Parks, National Monuments, National Forests, Wilderness areas, and other areas.

The land area of the contiguous United States is approximately 1.9 billion acres (770 million hectares). Alaska, separated from the contiguous United States by Canada, is the largest state at 365 million acres (150 million hectares). Hawaii, occupying an archipelago in the central Pacific, southwest of North America, has just over 4 million acres (1.6 million hectares). After Russia and Canada, the United States is the world's third or fourth largest nation by total area (land and water), ranking just above or below China. The ranking varies depending on how two territories disputed by China and India are counted and how the total size of the United States is calculated: the CIA World Factbook gives 3,794,101 square miles (9,826,675 km^2), the United Nations Statistics Division gives 3,717,813 sq mi (9,629,091 km^2), and the Encyclopædia Britannica gives 3,676,486 sq mi (9,522,055 km^2). Including only land area, the United States is third in size behind Russia and China, just ahead of Canada.

The coastal plain of the Atlantic seaboard gives way further inland to deciduous forests and the rolling hills of the Piedmont. The Appalachian Mountains divide the eastern seaboard from the Great Lakes and the grasslands of the Midwest. The Mississippi–Missouri River, the world's fourth longest river system, runs mainly north–south through the heart of the country. The flat, fertile prairie of the Great Plains stretches to the west, interrupted by a highland region in the southeast. The Rocky Mountains, at the western edge of the Great Plains, extend north to south across the country, reaching altitudes higher than 14,000 feet (4,300 m) in Colorado. Farther west are the rocky Great Basin and deserts such as the Mojave. The Sierra Nevada and Cascade mountain ranges run close to the

Pacific coast. At 20,320 feet (6,194 m), Alaska's Mount McKinley is the tallest peak in the country and in North America. Active volcanoes are common throughout Alaska's Alexander and Aleutian Islands, and Hawaii consists of volcanic islands. The supervolcano underlying Yellowstone National Park in the Rockies is the continent's largest volcanic feature.

The United States, with its large size and geographic variety, includes most climate types. To the east of the 100th meridian, the climate ranges from humid continental in the north to humid subtropical in the south. The southern tip of Florida is tropical, as is Hawaii. The Great Plains west of the 100th meridian are semi-arid. Much of the Western mountains are alpine. The climate is arid in the Great Basin, desert in the Southwest, Mediterranean in coastal California, and oceanic in coastal Oregon and Washington and southern Alaska. Most of Alaska is subarctic or polar. Extreme weather is not uncommon—the states bordering the Gulf of Mexico are prone to hurricanes, and most of the world's tornadoes occur within the country, mainly in the Midwest's Tornado Alley.

The U.S. ecology is considered "megadiverse": about 17,000 species of vascular plants occur in the contiguous United States and Alaska, and over 1,800 species of flowering plants are found in Hawaii, few of which occur on the mainland. The United States is home to more than 400 mammal, 750 bird, and 500 reptile and amphibian species. About 91,000 insect species have been described. The Endangered Species Act of 1973 protects threatened and endangered species and their habitats, which are monitored by the United States Fish and Wildlife Service. There are fifty-eight national parks and hundreds of other federally managed parks, forests, and wilderness areas. Altogether, the government owns 28.8% of the country's land area. Most of this is protected, though some is leased for oil and gas drilling, mining, logging, or cattle ranching; 2.4% is used for military purposes.

Native Americans and European Settlers

The indigenous peoples of the U.S. mainland, including Alaska Natives, are most commonly believed to have migrated from Asia. They began arriving at least 12,000 and as many as 40,000 years ago. Some, such as the pre-Columbian Mississippian culture, developed advanced agriculture, grand architecture, and state-level societies. After Europeans began settling the Americas, many millions of indigenous Americans died from epidemics of imported diseases such as smallpox.

In 1492, Genoese explorer Christopher Columbus, under contract to the Spanish crown, reached several Caribbean islands, making first contact with the indigenous people. On April 2, 1513, Spanish conquistador Juan Ponce de Leon landed on what he called "La Florida"—the first documented European arrival on what would become the U.S. mainland. Spanish settlements in the region were followed by ones in the present-day southwestern United States that drew thousands through Mexico. French fur traders established outposts of New France around the Great Lakes; France eventually claimed much of the North American interior, down to the Gulf of Mexico. The first successful English settlements were the Virginia Colony in Jamestown in 1607 and the Pilgrims' Plymouth Colony in 1620. The 1628 chartering of the Massachusetts Bay Colony resulted in a wave of migration; by 1634, New England had been settled by some 10,000 Puritans. Between the late 1610s and the American Revolution, about 50,000 convicts were shipped to Britain's American colonies. Beginning in 1614, the Dutch settled along the lower Hudson River, including New Amsterdam on Manhattan Island.

In 1674, the Dutch ceded their American territory to England; the province of New Netherland was renamed New York. Many new immigrants, especially to the South, were indentured servants—some two-thirds of all Virginia immigrants between 1630 and 1680. By the turn of the century, African slaves were becoming the primary source of bonded labour. With the 1729 division of the Carolinas and the 1732 colonization of Georgia, the thirteen British colonies that would become the United States of America were established. All had local governments with elections open to most free men, with a growing devotion to the ancient rights of Englishmen and a sense of self-government stimulating support for republicanism. All legalized the African slave trade. With high birth rates, low death rates, and steady immigration, the colonial population grew rapidly. The Christian revivalist movement of the 1730s and 1740s known as the Great Awakening fuelled interest in both religion and religious liberty. In the French and Indian War, British forces seized Canada from the French, but the francophone population remained politically isolated from the southern colonies. Excluding the Native Americans (popularly known as "American Indians"), who were being displaced, those thirteen colonies had a population of 2.6 million in 1770, about one-third that of Britain; nearly one in five Americans were black slaves. Though subject to British taxation, the American colonials had no representation in the Parliament of Great Britain.

Independence and Expansion

Tensions between American colonials and the British during the revolutionary period of the 1760s and early 1770s led to the American Revolutionary War, fought from 1775 through 1781. On June 14, 1775, the Continental Congress, convening in Philadelphia, established a Continental Army under the command of George Washington. Proclaiming that "all men are created equal" and endowed with "certain unalienable Rights," the Congress adopted the Declaration of Independence, drafted largely by Thomas Jefferson, on July 4, 1776. That date is now celebrated annually as America's Independence Day. In 1777, the Articles of Confederation established a weak confederal government that operated until 1789.

After the British defeat by American forces assisted by the French, Great Britain recognized the independence of the United States and the states' sovereignty over American territory west to the Mississippi River. A constitutional convention was organized in 1787 by those wishing to establish a strong national government, with powers of taxation. The United States Constitution was ratified in 1788, and the new republic's first Senate, House of Representatives, and president—George Washington—took office in 1789. The Bill of Rights, forbidding federal restriction of personal freedoms and guaranteeing a range of legal protections, was adopted in 1791.

Attitudes toward slavery were shifting; a clause in the Constitution protected the African slave trade only until 1808.

The Northern states abolished slavery between 1780 and 1804, leaving the slave states of the South as defenders of the "peculiar institution." The Second Great Awakening, beginning about 1800, made evangelicalism a force behind various social reform movements, including abolitionism.

Americans' eagerness to expand westward prompted a long series of Indian Wars. The Louisiana Purchase of French-claimed territory under President Thomas Jefferson in 1803 almost doubled the nation's size. The War of 1812, declared against Britain over various grievances and fought to a draw, strengthened U.S. nationalism. A series of U.S. military incursions into Florida led Spain to cede it and other Gulf Coast territory in 1819. The Trail of Tears in the 1830s exemplified the Indian removal policy that stripped the native peoples of their land. The United States annexed the Republic of Texas in 1845. The concept of Manifest Destiny was popularized during this time. The 1846 Oregon Treaty with Britain led to U.S. Control of the present-day American Northwest. The U.S. Victory in the Mexican–American War resulted in the 1848 cession of California and much of the present-day American Southwest. The California Gold Rush of 1848–49 further spurred western migration. New railways made relocation easier for settlers and increased conflicts with Native Americans. Over a half-century, up to 40 million American bison, or buffalo, were slaughtered for skins and meat and to ease the railways' spread. The loss of the buffalo, a primary resource for the plains Indians, was an existential blow to many native cultures.

Civil War and Industrialization

Tensions between slave and free states mounted with arguments over the relationship between the state and federal governments, as well as violent conflicts over the spread of slavery into new states. Abraham Lincoln, candidate of the largely antislavery Republican Party, was elected president in 1860. Before he took office, seven slave states declared their secession—which the federal government maintained was illegal—and formed the Confederate States of America. With the Confederate attack upon Fort Sumter, the American Civil War began and four more slave states joined the Confederacy.

Lincoln's Emancipation Proclamation in 1863 declared slaves in the Confederacy to be free. Following the Union victory in 1865, three amendments to the U.S. Constitution ensured freedom for the nearly four million African Americans who had been slaves, made them citizens, and gave them voting rights. The war and its resolution led to a substantial increase in federal power.

After the war, the assassination of Lincoln radicalized Republican Reconstruction policies aimed at reintegrating and rebuilding the Southern states while ensuring the rights of the newly freed slaves. The resolution of the disputed 1876 presidential election by the Compromise of 1877 ended Reconstruction; Jim Crow laws soon disenfranchised many African Americans. In the North, urbanization and an unprecedented influx of immigrants from Southern and Eastern Europe hastened the country's industrialization. The wave of immigration, lasting until 1929, provided labour and transformed American culture. National infrastructure development spurred economic growth. The 1867 Alaska purchase from Russia completed the country's mainland expansion. The Wounded Knee massacre in 1890 was the last major armed conflict of the Indian Wars. In 1893, the indigenous monarchy of the Pacific Kingdom of Hawaii was overthrown in a coup led by American residents; the United States annexed the archipelago in 1898. Victory in the Spanish–American War the same year demonstrated that the United States was a world power and led to the annexation of Puerto Rico, Guam, and the Philippines. The Philippines gained independence a half-century later; Puerto Rico and Guam remain U.S. Territories.

World War I, Great Depression, and World War II

At the outbreak of World War I in 1914, the United States remained neutral. Most Americans sympathized with the British and French, although many opposed intervention. In 1917, the United States joined the Allies, helping to turn the tide against the Central Powers. After the war, the Senate did not ratify the Treaty of Versailles, which established the League of Nations. The country pursued a policy of unilateralism, verging on isolationism. In 1920, the women's rights movement won passage of a constitutional amendment granting women's suffrage. The prosperity of the Roaring Twenties ended with the Wall Street Crash of 1929 that triggered the Great Depression. After his election as president in 1932, Franklin D. Roosevelt responded with the New Deal, a range of policies increasing government intervention in the economy. The Dust Bowl of the mid-1930s impoverished many farming communities and spurred a new wave of western migration.

The United States, effectively neutral during World War II's early stages after Nazi Germany's invasion of Poland in September 1939, began supplying materiel to the Allies in March 1941 through the Lend-Lease program. On December 7, 1941, the Empire of Japan launched a surprise attack on Pearl Harbour, prompting the United States to join the Allies against the Axis powers as well as the internment of Japanese Americans by the thousands. Participation in the war spurred capital investment and industrial capacity. Among the major combatants, the United States was the only nation to become richer—indeed, far richer—instead of poorer because of the war. Allied conferences at Bretton Woods and Yalta outlined a new system of international organizations that placed the United States and Soviet Union at the centre of world affairs. As victory was won in Europe, a 1945 international conference held in San Francisco produced the United Nations Charter, which became active after the war. The United States, having developed the first nuclear weapons, used them on the Japanese cities of Hiroshima and Nagasaki in August. Japan surrendered on September 2, ending the war.

Cold War and Protest Politics

The United States and Soviet Union jockeyed for power after World War II during the Cold War, dominating the military affairs of Europe through NATO and the Warsaw Pact. The United States promoted liberal democracy and capitalism, while the Soviet Union promoted communism and a centrally planned economy. Both supported dictatorships and engaged in proxy wars. American troops fought Communist Chinese forces in the Korean War of 1950–53. The House Un-American Activities

Committee pursued a series of investigations into suspected leftist subversion, while Senator Joseph McCarthy became the urehead of anticommunist sentiment.

The 1961 Soviet launch of the first manned spaceflight prompted President John F. Kennedy's call for the United States to be first to land "a man on the moon," achieved in 1969. Kennedy also faced a tense nuclear showdown with Soviet forces in Cuba. Meanwhile, the United States experienced sustained economic expansion. A growing civil rights movement, symbolized and led by African Americans such as Rosa Parks and Martin Luther King, Jr. used nonviolence to confront segregation and discrimination. Following Kennedy's assassination in 1963, the Civil Rights Act of 1964 and Voting Rights Act of 1965 were passed under President Lyndon B. Johnson. Johnson and his successor, Richard Nixon, expanded a proxy war in Southeast Asia into the unsuccessful Vietnam War. A widespread countercultural movement grew, fuelled by opposition to the war, black nationalism, and the sexual revolution. Betty Friedan, Gloria Steinem, and others led a new wave of feminism that sought political, social, and economic equality for women.

As a result of the Watergate scandal, in 1974 Nixon became the first U.S. President to resign, to avoid being impeached on charges including obstruction of justice and abuse of power; he was succeeded by Vice President Gerald Ford. The Jimmy Carter administration of the late 1970s was marked by stagflation and the Iran hostage crisis. The election of Ronald Reagan as president in 1980 heralded a rightward shift in American politics, reflected in major changes in taxation and spending priorities. His second term in office brought both the Iran-Contra scandal and significant diplomatic progress with the Soviet Union. The subsequent Soviet collapse ended the Cold War.

Contemporary Era

Under President George H. W. Bush, the United States took a lead role in the UN–sanctioned Gulf War. The longest economic expansion in modern U.S. History—from March 1991 to March 2001—encompassed the Bill Clinton administration and the dot-com bubble. A civil lawsuit and sex scandal led to Clinton's impeachment in 1998, but he remained in office. The 2000 presidential election, one of the closest in American history, was resolved by a U.S. Supreme Court decision—George W. Bush, son of George H. W. Bush, became president.

On September 11, 2001, al-Qaeda terrorists struck the World Trade Centre in New York City and The Pentagon near Washington, D.C., killing nearly three thousand people. In response, the Bush administration launched a "War on Terrorism". In late 2001, U.S. Forces led an invasion of Afghanistan, removing the Taliban government and al-Qaeda training camps. Taliban insurgents continue to fight a guerrilla war. In 2002, the Bush administration began to press for regime change in Iraq on controversial grounds. Lacking the support of NATO or an explicit UN mandate for military intervention, Bush organized a Coalition of the Willing; coalition forces preemptively invaded Iraq in 2003, removing dictator Saddam Hussein. In 2005, Hurricane Katrina caused severe destruction along much of the Gulf Coast, devastating New Orleans. On November 4, 2008, amid a global economic recession, Barack Obama was elected president. He is the first African American to hold the office. In early 2010, he oversaw the enactment of major health care reform.

Government and Elections

The United States is the world's oldest surviving federation. It is a constitutional republic and representative democracy, "in which majority rule is tempered by minority rights protected by law." The government is regulated by a system of checks and balances defined by the U.S. Constitution, which serves as the country's supreme legal document. In the American federalist system, citizens are usually subject to three levels of government, federal, state, and local; the local government's duties are commonly split between county and municipal governments. In almost all cases, executive and legislative officials are elected by a plurality vote of citizens by district. There is no proportional representation at the federal level, and it is very rare at lower levels. The federal government is composed of three branches:

- Legislative: The bicameral Congress, made up of the Senate and the House of Representatives, makes federal law, declares war, approves treaties, has the power of the purse, and has the power of impeachment, by which it can remove sitting members of the government.
- Executive: The president is the commander-in-chief of the military, can veto legislative bills before they become law, and appoints the members of the Cabinet (subject to Senate approval) and other officers, who administer and enforce federal laws and policies.
- Judicial: The Supreme Court and lower federal courts, whose judges are appointed by the president with Senate approval, interpret laws and overturn those they find unconstitutional.

The House of Representatives has 435 voting members, each representing a congressional district for a two-year term. House seats are apportioned among the states by population every tenth year. As of the 2000 census, seven states have the minimum of one representative, while California, the most populous state, has fifty-three. The Senate has 100 members with each state having two senators, elected at-large to six-year terms; one third of Senate seats are up for election every other year. The president serves a four-year term and may be elected to the office no more than twice. The president is not elected by direct vote, but by an indirect electoral college system in which the determining votes are apportioned by state. The Supreme Court, led by the Chief Justice of the United States, has nine members, who serve for life.

The state governments are structured in roughly similar fashion; Nebraska uniquely has a unicameral legislature. The governor (chief executive) of each state is directly elected. Some state judges and cabinet officers are appointed by the governors of the respective states, while others are elected by popular vote.

All laws and governmental procedures are subject to judicial review, and any law ruled in violation of the Constitution is voided. The original text of the Constitution establishes the structure and responsibilities of the federal government and its relationship with the individual states. Article One protects the right to the "great writ" of habeas corpus, and Article Three guarantees the right to a jury trial in all criminal cases. Amendments to the Constitution require the approval of three-fourths of the states. The Constitution has been amended twenty-seven times; the first ten amendments, which make up the Bill of Rights, and the Fourteenth Amendment form the central basis of Americans' individual rights.

Parties, Ideology and Politics

The United States has operated under a two-party system for most of its history. For elective offices at all levels, state-administered primary elections choose the major party nominees for subsequent general elections. Since the general election of 1856, the major parties have been the Democratic Party, founded in 1824, and the Republican Party, founded in 1854. Since the Civil War, only one third-party presidential candidate—former president Theodore Roosevelt, running as a Progressive in 1912—has won as much as 20% of the popular vote.

Within American political culture, the Republican Party is considered centre-right or "conservative" and the Democratic Party is considered centre-left or "liberal". The states of the Northeast and West Coast and some of the Great Lakes states, known as "blue states", are relatively liberal. The "red states" of the South and parts of the Great Plains and Rocky Mountains are relatively conservative. The winner of the 2008 presidential election, Democrat Barack Obama, is the 44th U.S. President. All previous presidents were men of solely European descent. The 2008 elections also saw the Democratic Party strengthen its control of both the House and the Senate. In the 111th United States Congress, the Senate comprises 57 Democrats, two independents who caucus with the Democrats, and 41 Republicans; the House comprises 255 Democrats and 177 Republicans (three seats are vacant). There are 26 Democratic and 24 Republican state governors.

Political Divisions

The United States is a federal union of fifty states. The original thirteen states were the successors of the thirteen colonies that rebelled against British rule. Early in the country's history, three new states were organized on territory separated from the claims of the existing states: Kentucky from Virginia; Tennessee from North Carolina; and Maine from Massachusetts. Most of the other states have been carved from territories obtained through war or purchase by the U.S. Government. One set of exceptions comprises Vermont, Texas, and Hawaii: each was an independent republic before joining the union. During the American Civil War, West Virginia broke away from Virginia. The most recent state—Hawaii—achieved statehood on August 21, 1959. The states do not have the right to secede from the union.

The states compose the vast bulk of the U.S. Land mass; the two other areas considered integral parts of the country are the District of Columbia, the federal district where the capital, Washington, is located; and Palmyra Atoll, an uninhabited but incorporated territory in the Pacific Ocean. The United States also possesses five major overseas territories: Puerto Rico and the United States Virgin Islands in the Caribbean; and American Samoa, Guam, and the Northern Mariana Islands in the Pacific. Those born in the territories (except for American Samoa) possess U.S. Citizenship. American citizens residing in the territories have many of the same rights and responsibilities as citizens residing in the states; however, they are generally exempt from federal income tax, may not vote for president, and have only nonvoting representation in the U.S. Congress.

Foreign Relations and Military

The United States exercises global economic, political, and military influence. It is a permanent member of the United Nations Security Council and New York City hosts the United Nations Headquarters. It is a member of the G8, G20, and Organisation for Economic Co-operation and Development. Almost all countries have embassies in Washington, D.C., and many have consulates around the country. Likewise, nearly all nations host American diplomatic missions. However, Cuba, Iran, North Korea, Bhutan, Sudan, and the Republic of China (Taiwan) do not have formal diplomatic relations with the United States.

The United States enjoys strong ties with the United Kingdom, Canada, Australia, New Zealand, Japan, South Korea, and Israel. It works closely with fellow NATO members on military and security issues and with its neighbours through the Organization of American States and free trade agreements such as the trilateral North American Free Trade Agreement with Canada and Mexico. In 2008, the United States spent a net $25.4 billion on official development assistance, the most in the world. As a share of gross national income (GNI), however, the U.S. Contribution of 0.18% ranked last among twenty-two donor states. In contrast, private overseas giving by Americans is relatively generous.

The president holds the title of commander-in-chief of the nation's armed forces and appoints its leaders, the secretary of defence and the Joint Chiefs of Staff. The United States Department of Defence administers the armed forces, including the Army, Navy, Marine Corps, and Air Force. The Coast Guard is run by the Department of Homeland Security in peacetime and the Department of the Navy in time of war. In 2008, the armed forces had 1.4 million personnel on active duty. The Reserves and National Guard brought the total number of troops to 2.3 million. The Department of Defence also employed about 700,000 civilians, not including contractors.

Military service is voluntary, though conscription may occur in wartime through the Selective Service System. American forces can be rapidly deployed by the Air Force's large fleet of transport aircraft, the Navy's eleven active aircraft carriers, and Marine Expeditionary Units at sea with the Navy's Atlantic and Pacific fleets. The military operates 865 bases and facilities abroad, and maintains deployments greater than 100 active duty personnel in 28 foreign countries. The extent of this global military presence has prompted some scholars to describe the United States as maintaining an "empire of bases."

Total U.S. Military spending in 2008, more than $600 billion, was over 41% of global military spending and greater than the next fourteen largest national military expenditures combined. The per capita spending of $1,967 was about nine times the world average; at 4% of GDP, the rate was the second-highest among the top fifteen military spenders, after Saudi Arabia.

The proposed base Department of Defence budget for 2010, $533.8 billion, is a 4% increase over 2009 and 80% higher than in 2001; an additional $130 billion is proposed for the military campaigns in Iraq and Afghanistan. As of May 2010, there were 94,000 U.S. Troops deployed to Afghanistan, and 92,000 to Iraq. As of June 2, 2010, the United States had suffered 4,400 military fatalities during the Iraq War, and 1,087 during the War in Afghanistan.

Economy

The United States has a capitalist mixed economy, which is fuelled by abundant natural resources, a well-developed infrastructure, and high productivity. According to the International Monetary Fund, the U.S. GDP of $14.4 trillion constitutes 24% of the gross world product at market exchange rates and almost 21% of the gross world product at purchasing power parity (PPP). It has the largest national GDP in the world, though it is about 5% less than the GDP of the European Union at PPP in 2008. The country ranks seventeenth in the world in nominal GDP per capita and sixth in GDP per capita at PPP.

The United States is the largest importer of goods and third largest exporter, though exports per capita are relatively low. In 2008, the total U.S. Trade deficit was $696 billion. Canada, China, Mexico, Japan, and Germany are its top trading partners. In 2007, vehicles constituted both the leading import and leading export commodity. Japan is the largest foreign holder of U.S. Public debt, having surpassed China in early 2010. The United States ranks second in the Global Competitiveness Report.

In 2009, the private sector is estimated to constitute 55.3% of the economy, with federal government activity accounting for 24.1% and state and local government activity (including federal transfers) the remaining 20.6%. The economy is postindustrial, with the service sector contributing 67.8% of GDP, though the United States remains an industrial power. The leading business field by gross business receipts is wholesale and retail trade; by net income it is manufacturing. Chemical products are the leading manufacturing field. The United States is the third largest producer of oil in the world, as well as its largest importer. It is the world's number one producer of electrical and nuclear energy, as well as liquid natural gas, sulfur, phosphates, and salt. While agriculture accounts for just under 1% of GDP, the United States is the world's top producer of corn and soybeans. The New York Stock Exchange is the world's largest by dollar volume. Coca-Cola and McDonald's are the two most recognized brands in the world.

In the third quarter of 2009, the American labour force comprised 154.4 million people. Of those employed, 81% had jobs in the service sector. With 22.4 million people, government is the leading field of employment. About 12% of workers are unionized, compared to 30% in Western Europe. The World Bank ranks the United States first in the ease of hiring and firing workers. Between 1973 and 2003, a year's work for the average American grew by 199 hours. Partly as a result, the United States maintains the highest labour productivity in the world. In 2008, it also led the world in productivity per hour, overtaking Norway, France, Belgium and Luxembourg, which had surpassed the United States for most of the preceding decade. Compared to Europe, U.S. Property and corporate income tax rates are generally higher, while labour and, particularly, consumption tax rates are lower.

Income and Human Development

According to the United States Census Bureau, the pretax median household income in 2007 was $50,233. The median ranged from $68,080 in Maryland to $36,338 in Mississippi. Using purchasing power parity exchange rates, the overall median is similar to the most affluent cluster of developed nations. After declining sharply during the middle of the 20th century, poverty rates have plateaued since the early 1970s, with 11–15% of Americans below the poverty line every year, and 58.5% spending at least one year in poverty between the ages of 25 and 75. In 2007, 37.3 million Americans lived in poverty.

The U.S. welfare state is now among the most austere in the developed world, reducing both relative poverty and absolute poverty by considerably less than the mean for rich nations, though combined private and public social expenditures per capita are higher than in any of the Nordic countries. While the American welfare state does well in reducing poverty among the elderly, the young receive relatively little assistance. A 2007 UNICEF study of children's well-being in twenty-one industrialized nations ranked the United States next to last.

Despite strong increases in productivity, low unemployment, and low inflation, income gains since 1980 have been slower than in previous decades, less widely shared, and accompanied by increased economic insecurity. Between 1947 and 1979, real median income rose by over 80% for all classes, with the incomes of poor Americans rising faster than those of the rich. Median household income has increased for all classes since 1980, largely owing to more dual-earner households, the closing of the gender gap, and longer work hours, but growth has been slower and strongly tilted toward the very top. Consequently, the share of income of the top 1%—21.8% of total reported income in 2005—has more than doubled since 1980, leaving the United States with the greatest income inequality among developed nations. The top 1% pays 27.6% of all federal taxes; the top 10% pays 54.7%. Wealth, like income, is highly concentrated: The richest 10% of the adult population possesses 69.8% of the country's household wealth, the second-highest share among developed nations. The top 1% possesses 33.4% of net wealth.

Science and Technology

The United States has been a leader in scientific research and technological innovation since the late 19th century. In 1876, Alexander Graham Bell was awarded the first U.S. Patent for the telephone. Thomas Edison's laboratory developed the phonograph, the first long-lasting light bulb, and the first viable movie camera. Nikola Tesla pioneered alternating current, the AC motor, and radio. In the early 20th century, the automobile companies of Ransom E. Olds and Henry Ford promoted the assembly line. The Wright brothers, in 1903, made the first sustained and controlled heavier-than-air powered flight.

The rise of Nazism in the 1930s led many European scientists, including Albert Einstein and Enrico Fermi, to immigrate to the United States. During World War II, the Manhattan Project developed nuclear weapons, ushering in the Atomic Age. The Space Race produced rapid advances in rocketry, materials science, and computers. The United States largely developed the ARPANET and its successor, the Internet. Today, the bulk of research and development funding, 64%, comes from the private sector. The United States leads the world in scientific research papers and impact factor. Americans possess high levels of technological consumer goods, and almost half of U.S. households have broadband Internet access. The country is the primary developer and grower of genetically modified food; more than half of the world's land planted with biotech crops is in the United States.

Transportation

Transportation in the United States is facilitated by road, air, rail, and water networks. The vast majority of passenger travel occurs by automobile for shorter distances, and airplane for longer distances. In descending order, most cargoes travel by railroad, truck, pipeline, or boat; air shipping is typically used only for perishables and premium express shipments.

Ownership and Jurisdiction

The overwhelming majority of roads in the United States are owned and maintained by state and local governments. Federally maintained roads are generally found only on federal lands (such as national parks) and at federal facilities (like military bases). The Interstate Highway System is funded by the federal government but maintained by state governments. There are a few private highways in the United States, which use tolls to pay for construction and maintenance. There are many local private roads, generally serving remote or insular residences.

Passenger and freight rail systems, bus systems, water ferries, and dams may be under either public or private ownership and operation. Civilian airlines are all privately owned and financed. Most airports are owned and operated by local government authorities, but there are also some private airports. The Transportation Security Administration has provided security at most major airports since 2001.

The United States Department of Transportation and its divisions provide regulation, supervision, and funding for all aspects of transportation, except for customs, immigration, and security, which are the responsibility of the United States Department of Homeland Security. Each state has its own Department of Transportation, which builds and maintains state highways, and depending upon the state, may either directly operate or supervise other modes of transportation.

Aviation law is almost entirely a federal matter, while automobile traffic laws are enacted and enforced by state and local authorities. Economic jurisdiction over tidelands is shared between the state and federal governments, while the United States Coast Guard is the primary enforcer of law and security on US waterways.

Mode Share

Passenger

Passenger transportation is dominated by a network of over 3.9 million miles of highways which is pervasive and highly developed by global standards. Passenger transportation is dominated by passenger vehicles (including cars, trucks, vans, and motorcycles), which account for 86% of passenger-miles travelled. The remaining 14% was handled by planes, trains, and buses.

As of 2003, there were 759 automobiles per 1,000 Americans, compared to 472 per 1,000 inhabitants of the European Union the following year. The US intercity passenger rail system is limited in scope. Bicycle usage is minimal.

Cargo

Freight transportation is carried by a variety of networks. The largest percentage of US freight is carried by trucks (60%), followed by pipelines (18%), rail (10%), ship (8%), and air (0.01%). Other modes of transportation, such as parcels and intermodal freight accounted for about 3% of the remainder. Air freight is commonly used only for perishables and premium express shipments. The difference in percentage of rail's share by ton-miles and by weight (10% vs 38%) is accounted for by the extreme efficiency of trains. A single railroad locomotive may pull fifty boxcars full of freight while a truck only pulls one. Trucks surpass trains in the weight category due their greater numbers, while trains surpass trucks in the ton-miles category due to the vast distances they travel carrying large amounts of freight.

Usually cargo, apart from petroleum and other bulk commodities, is imported in containers through seaports, then distributed by road and rail. The quasi-governmental United States Postal Service has a monopoly on letter delivery (except for express services) but several large private companies such as FedEx and UPS compete in the package and cargo delivery market.

Safety

U.S. passenger fatalities per billion passenger-miles, 2002-2007			
Mode	**Passenger fatalities**	**Passengermiles (millions)**	**Passenger fatalities per billion passenger miles**
Passenger car	127,124	15,958,620 M	7.97
Light rail	79	9,980 M	7.92
Motor bus	399	117,982 M	3.38
Commuter rail	105	59,736 M	1.76
Heavy rail (subway)	106	86,900 M	1.22
Railroad (intercity)	36	33,234 M	1.08
Airline	113	3,326,286 M	0.03

History

Speed to destination is an important factor in choosing a mode of transportation. In the late 18th century overland transportation was by horse, while water and river transportation was primarily by sailing vessel. The United States population was centered on its Atlantic coast, with all major population centers located on a natural harbour or navigable waterway. Low population density between these centers resulted in a heavily reliance on coastwise and riverboat shipping. The first government expenditures on highway transportation were funded to speed the delivery of overland mail, such as the Boston Post Road between New York City and Boston. Due to the distances between these population centers and the cost to maintain the roads, many highways in the late 18th century and early 19th century were private turnpikes. Other highways were mainly unimproved and impassable by wagon at least some of the year. Economic expansion in the late 18th century to early 19th century spurred the building of canals to speed goods to market, of which the most prominently successful example was the Erie Canal.

Development of the mid-western and southern states drained by the Mississippi River system (Mississippi, Ohio and Missouri Rivers) was accelerated by the introduction of steamboats on these rivers in the early 19th Century. Prior to the introduction of steamboats, transit upstream was impractical because of strong currents on parts of these waterways. Steamboats provided both passenger and freight transportation until the development of railroads later in the 19th Century gradually reduced their presence.

The rapid expansion of Railroads brought the canal boom to a sudden end, providing a quick, scheduled and year round mode of transportation that quickly spread to interconnect the states by the mid-19th century. During the industrialization of the United States after the Civil War, railroads, led by the transcontinental rail system in the 1860s, expanded quickly across the United States to serve industries and the growing cities. During the late 19th century, railroads often had built redundant routes to a competitor's road or built through sparsely populated regions that generated little traffic. These marginal rail routes survived the pricing pressures of competition, or the lack of revenue generated by low traffic, as long as railroads provided the only efficient economical way to move goods and people across the United States. In addition to the intercity passenger network running on Class I and II railroads, a large network of interurban (trolley or "street running") rail lines extended out from the cities and interchanged passenger and freight traffic with the railroads and also provided competition.

The advent of the automobile signaled the end of railroads as the predominant transportation for people and began an era of mobility in the United States that added greatly to its economic output. The early 20th century Lincoln Highway and other auto trails gave way in the 1920s to an early national highway system making the automobile the preferred mode of travel for most Americans. Interurban rail service declined, followed by trolley cars due in part to the advent of motorized buses and the lack of dedicated rights-of-way but also by deliberate efforts to dismantle urban rail infrastructure. The scarcity of industrial materials during World War Two slowed the growth of the automobile, briefly reemphasizing much of the nation's declining rail network. In the 1950s, the United States renewed building a network of high-capacity, high-speed highways to link its vast territory. The most important element is the Interstate Highway system, first commissioned in the 1950s by President Dwight D. Eisenhower and modelled after the Italian autostrada and the German Autobahn system.

With the advent of commercial airline industry, intercity rail began to suffer a loss of ridership. As the civil air transportation network of airports and other infrastructure expanded, air travel became more accessible to the general population. Technological advances ushered in the jet age, which increased airline capacity, while decreasing travel times and the cost of flights. The costs of flying rapidly decreased intercity rail ridership by the late 1960s to a point where railroads could no longer profitably operate networks of passenger trains. By the early 1970s almost all passenger rail operation and ownership had been transferred to various federal, municipal and state agencies. Passenger rail came to be heavily subsidized, as it is today.

Freight railroads continued to decline as motor freight captured a significant portion of the less-than-carload business. This loss of business, when combined the highly regulated operating environment and constrained pricing power, forced many railroads into receivership and the nationalization of several critical eastern carriers into the Consolidated Rail Corporation (Conrail). Deregulation of the railroads by the Staggers Act in 1980 created a regulatory environment more favourable to the economics of the railroad industry. In the 1990s, the increase in foreign trade and intermodal container shipping led to a revival of the freight railroads, which have effectively consolidated into two eastern and two western private transportation networks: Union Pacific and BNSF in the west, and CSX and Norfolk Southern in the east. Canadian National Railway took over the Illinois Central route down the Misssippi River valley.

Wartime expediency encouraged long distance pipeline transport of petroleum and natural gas, which was greatly expanded in the middle 20th century to take over most of the domestic long-haul market.

Road Transportation

Infrastructure and Private Automobile Use

In comparison to most of the Western world, the United States relies much more heavily on its roads both for commercial and personal transit. Car ownership is nearly universal except in few of the largest cities where extensive mass transit systems have been built.

With the development of the extensive Eisenhower Interstate Highway System in the 1950s, both long-distance trips and daily the commute were mostly by private automobile. This network was designed to exacting federal standards in order to receive federal funding. The system, as of 2004, has a total length of 46,837 miles (75,376 km), making it both the largest expressway system in the world and the largest public works project in US history.

The Interstate system joined an existing National Highway System, comprising 160,000 miles (256,000 kilometres) of roadway, a fraction of the total mileage of roads. The Interstate system serves nearly all major U.S. Cities, often through the downtown areas (a point which triggered freeway and expressway revolts in the 1960s and 1970s). The distribution of virtually all goods and services involves Interstate highways at some point. Residents of American cities commonly use urban Interstates to travel to their places of work. The vast majority of long-distance travel, whether for vacation or business, is by the national road network; of these trips, about one-third (by the total number of miles driven in the country in 2003) utilize the Interstate system.

In addition to the routes of the Interstate system, there are those of the US Highway system, not to be confused with the above mentioned National Highway System. These networks are further supplemented by State Highways, and the local roads of counties, municipal streets, and federal agencies, such as the Bureau of Indian Affairs. Altogether there are more than 4,209,835 km of paved roads in the U.S. (Including 75,040 km of expressways), and 2,255,964 km of unpaved roads.

All highways are maintained by state governments, although they receive federal aid to build and maintain freeways signed as part of the 46,000 mile (75,000 km) nationwide Interstate highway network. A large number of expressways are actually government or privately-operated toll roads in many East Coast and Midwestern states. West Coast freeways are generally free to users ("freeways", no toll charged per use), although since the 1990s there have been some small experiments with toll roads operated by private companies.

After the collapse of the I-35W Mississippi River bridge in Minnesota in August 2007, the backlog of road and bridge maintenance across the country became a prominent political issue.

As of 2010, seatbelt use is mandatory in all states except New Hampshire.

Intercity Bus

Greyhound Lines is the largest intercity bus company in the United States, with routes in all parts of the continental U.S.. There are also many smaller regional bus companies, many of which use the terminal and booking facilities provided by Greyhound. Intercity bus is, in most cases, the least expensive way to travel long distances in the US.

Congestion

Traffic congestion, especially at rush hour, is a problem in many of the country's larger cities. A 2009 study found that traffic congestion costs the US almost $87.2 billion. The economic costs of traffic congestion have increased 63% over the past decade, and despite declining traffic volumes, caused by the economic downturn, Americans still waste more than 2.8 billion gallons of fuel each year as a result of traffic congestion. Motorists also waste 4.2 billion hours annually, or one full work week per traveller.

Cargo

The trucking industry (also referred to as the transportation or logistics industry) involves the transport and distribution of commercial and industrial goods using commercial motor vehicles (CMV). In this case, CMVs are most often trucks; usually semi trucks, box trucks, or dump trucks. A truck driver (commonly referred to as a "trucker") is a person who earns a living as the driver of a CMV.

The trucking industry provides an essential service to the American economy by transporting large quantities of raw materials, works in process, and finished goods over land—typically from manufacturing plants to retail distribution centers. Trucks are also important to the construction industry, as dump trucks and portable concrete mixers are necessary to move the large amounts of rocks, dirt, concrete, and other building materials used in construction. Trucks in America are responsible for the majority of freight movement over land, and are vital tools in the manufacturing, transportation, and warehousing industries.

Large trucks and buses require a commercial driver's license (CDL) to operate. Obtaining a CDL requires extra education and training dealing with the special knowledge requirements and handling characteristics of such a large vehicle. Drivers of CMVs must adhere to the hours of service, which are regulations governing the driving hours of commercial drivers. These, and all other rules regarding the safety of interstate commercial driving, are issued by the Federal Motor Carrier Safety Administration (FMCSA). The FMCSA is also a division of the United States Department of Transportation (USDOT), which governs all transportation-related industries such as trucking, shipping, railroads, and airlines. Some other issues are handled by another branch of the USDOT, the Federal Highway Administration (FHWA).

Developments in technology, such as computers, satellite communication, and the internet, have contributed to many improvements within the industry. These developments have increased the productivity of company operations, saved the time and effort of drivers, and provided new, more accessible forms of entertainment to men and women who often spend long periods of time away from home. In 2006, the U.S. Environmental Protection Agency implemented revised emission standards for diesel trucks (reducing airborne pollutants emitted by diesel engines) which promises to improve air quality and public health.

Roadway links with adjacent countries and non-contiguous parts of the US:

- Canada-Yes.
- Mexico-Yes.
- Alaska-Yes, via Canada and the Alaska Marine Highway in Washington
- Russia-No, but proposed via Bering Strait crossing
- Cuba-No. Since the American embargo against Cuba, car ferry service from Florida and New Orleans to Havana ceased in 1962.

Air Transportation

The United States has advanced air transportation infrastructure which utilizes approximately 5,000 paved runways. In terms of passengers, seventeen of the world's thirty busiest airports in 2004 were in the U.S., including the world's busiest, Hartsfield-Jackson Atlanta International Airport. In terms of cargo, in the same year, twelve of the world's thirty busiest airports were in the U.S., including the world's busiest, Memphis International Airport. Private aircraft are also used for medical emergencies, government agencies, large businesses, and individuals).

Due to the geography of the United States and the generally large distances between major cities, air transportation is the preferred method of travel for trips over 300 miles, such as for business travellers and long distance vacation travellers. For cities closer together, such as Boston and New York City, New York and Washington D.C., and Philadelphia and D.C., air travel does not carry the majority of intercity traffic. There is no single national flag airline; passenger airlines in the United States have always been privately-owned. There are over 200 domestic passenger and cargo airlines and a number of international carriers. The international carriers of the U.S. are Delta Air Lines, American Airlines, United Airlines, Continental Airlines, and US Airways. There is currently no government regulation of ticket pricing, although the federal government retains jurisdiction over aircraft safety, pilot training, and accident investigations (through the Federal Aviation Administration and the National Transportation Safety Board). The Transportation Security Administration provides security at airports. With numerous airlines competing for traffic on the same intercity routes, ticket prices tend to be very competitive, resulting in low industry profit margins. This has led to frequent airline bankruptcies when ridership declines during economic recessions, jet fuel price increases, and other developments such as the September 11th terrorist attacks. Many airlines operate on a "hub and spoke" model. This system gives the predominant airline in a given airport a strong competitive position as it feeds passengers to and from the hub, maximizing the number of passengers on each flight. Examples of airline hubs include Delta Air Lines in Atlanta, US Airways in Charlotte or American Airlines in Dallas.

Public airports are usually constructed and operated by local governments. The main exceptions are on military bases. Like highways and passenger rail, the federal government subsidizes air travel with $14 billion of federal funds going to airport operations in 2002.

Air cargo comprises a large number of daily flights in the United States and are operated by private parcel companies such as FedEx and United Parcel Service. These organizations operate some of the largest fleets in the world. Most air cargo moved by these organizations is time sensitive overnight and 2nd day parcels. The U.S. Postal Service also moves much of its letters and time sensitive parcels via air, but on regularly scheduled passenger flights. At one time the U.S. Mail charged a premium for letters sent by airmail, but no longer does so except for overnight express mail.

Delays

Airline delays have been the subject of some controversy, prompting a GAO audit and Congressional debate in 2007-08.

Roughly one in four passengers experienced a passenger trip delay in 2007 and the average duration of delay experienced by these passengers was 1 hour 54 minutes. 24% of flights were delayed and 2% were cancelled entirely. Overall, passengers were delayed 320 million hours in 2007 with an estimated deadweight economic loss of up to $41 billion.

An average of 40% of passenger aircraft delays in the U.S. Originated in the New York metropolitan area, some in the area and others due to cascading effects. One-third of aircraft in the national airspace system move through the New York area at some point during a typical day.

To deal with delays, takeoff and landing scheduling caps have been imposed on certain urban airports at various times since 1968, including Washington Reagan National, Chicago O'Hare, and the three New York airports. Other short-term measures have been taken, including minor procedural changes, use of military airspace on peak travel days, and appointment of a "New York Airspace Czar" (Director for the New York Area Program Integration Office). The New York/New Jersey/ Philadelphia airspace is being "redesigned" incrementally, with completion estimated in 2012. The Bush Administration announced plans to auction some takeoff and landing slots at the New York airports, but plans were canceled by the Obama administration.

Longer-term solutions include increasing capacity by building more runways, and implementing the Next Generation Air Transportation System which would allow more direct flight paths.

Network Statistics:

- Airports: 14,951 (2008 est.)
- Airports with paved runways:
  - Total: 5,146
  - Over 10,000 ft: (3,047 m): 190

  - o 8,000 ft (2,438 m) to 10,000 ft (3,047 m): 227
  - o 5,000 ft (1,524 m) to 8,000 ft (2,437 m): 1,464
  - o 3,000 ft (914 m) to 5,000 ft (1,523 m): 2,307
  - o under 3,000 ft (914 m): 958 (1999 est.)
- Airports with unpaved runways:
  - o Total: 9,805
  - o 8,000 ft (2,438 m) to 10,000 ft (3,047 m): (2,437 m): 6
  - o 5,000 ft (1,524 m) to 8,000 ft (2,437 m): 156
  - o 3,000 ft (914 m) to 5,000 ft (1,523 m): 1,734
  - o Under 3,000 ft (914 m): 7,909 (1999 est.)
- Heliports: 146 (CIA, 2007 est.)

Rail

Passenger

Passenger trains were the dominant mode of transportation until the mid-twentieth century. The introduction of jet airplanes on major U.S. Routes and the completion of the Interstate Highway system accelerated a decline in intercity rail passenger demand during the 1960s, resulting in the sharp curtailment of passenger service by private railroads. This led to the creation of National Railroad Passenger Corporation (branded as Amtrak) by the Federal Government in 1971 to maintain limited intercity rail passenger service in most parts of the country. Amtrak serves most major cities but, outside of the Northeast, often by only one or two trains per day. Amtrak does not serve several major destinations, including Las Vegas, Nevada and Phoenix, Arizona. More frequent service is available in regional corridors between certain major cities, particularly the Northeast Corridor between Washington, D.C., New York City and Boston, between New York City and Albany, around Chicago, and in parts of California and the Pacific Northwest.

Cargo

The U.S. Makes extensive use of its rail system for freight. According to the Association of American Railroads: "U.S. freight railroads are the world's busiest, moving more freight than any rail system in any other country. In fact, U.S. railroads move more than four times as much freight as do all of Western Europe's freight railroads combined."

Nearly all railroad corridors (not including local transit rail systems) are owned by private companies that provide freight service. Amtrak pays these companies for the right to use the tracks for passenger service. There are approximately 150,000 mi (240,000 km) of mainline track in the United States — the world's longest national railroad network. See List of United States railroads

Freight rail transportation in Chicago is a major national bottleneck (about one-third of the nation's freight trains pass through the region), and the subject of a $1.5B infrastructure improvement project.

Railway links with adjacent countries:

- Canada-Yes. Amtrak connections run daily between New York City and Montreal, New York City and Toronto, and Seattle and Vancouver. Alaska is currently only rail-accessible by train ferry from Seattle. A proposed link to the Continental United States would link Alaska and Canada via the Yukon and British Columbia. The Canadian National Railway system includes the former Illinois Central route from Chicago via Memphis to New Orleans.
- Mexico-Yes. Several private firms run touristic trains from near El Paso, Texas through Chihuahua, Chihuahua and the Copper Canyon to El Fuerte, Sinaloa. Such trains also run from Nogales, Sonora, but no US passenger trains run near Nogales, Arizona, on the other side of the border. Another touristic train runs occasional trains between Campo, California, and Tecate, Baja California.
- Cuba-No. Since the United States embargo against Cuba, train ferry service from Florida and New Orleans to Havana ceased.
- Russia-No. But proposed tunnel or bridge under or over Bering Straits to connect with Trans-Siberian Railway.

With few exceptions, the rail gauge is standard gauge 4 ft 8+D $_2$ in (1,435 mm). The White Pass and Yukon Route from Skagway, Alaska to Whitehorse, Yukon by way of Bennett, British Columbia is 3 ft (914 mm) gauge.

Mass Transit

The miles travelled by passenger vehicles in the United States fell by 3.6% in 2008, while the number of trips taken on mass transit increased by 4.0%. At least part of the drop in urban driving can be explained by the 4% increase in the use of public transportation. Most medium-sized cities have some sort of local public transportation, usually a network of fixed bus routes. Larger cities tend to have metro rail systems (also known as heavy rail in the U.S.) And/or light rail systems for high-capacity passenger service within the urban area.

Legislation

On June 26, 2008, the House passed the Saving Energy Through Public Transportation Act (H.R. 6052), which gives grants to mass transit authorities to lower fares for commuters pinched at the pump and expand transit services. The bill also:

- Requires that all Federal agencies offer their employees transit pass transportation fringe benefits. Federal agencies within the National Capital Region have successful transit pass benefits programs.
- Increases the Federal cost-share of grants for construction of additional parking facilities at the end of subway lines from 80 to 100% to cover an increase in the number of people taking mass transit.
- Creates a pilot program for vanpool demonstration projects in urban and rural areas.
- Increases federal help for local governments to purchase alternative fuel buses, locomotives and ferries from 90 to 100%.

Water Transportation

Water transport is largely used for freight. Fishing and pleasure boats are numerous, and passenger service connects many of the nation's islands and remote coastal areas, crosses lakes, rivers, and harbours, and provides alternative access to Alaska which bypasses Canada. Several major seaports in the United States include New York to the east, Houston and New Orleans on the gulf coast, Los Angeles to the west. The interior of the U.S. also has major shipping channels, via the St. Lawrence Seaway and the Mississippi River system. Freight on the Mississippi River system is carried on barges pushed by approximately 8000 "towboats" and largely consists of bulk goods, such as petrochemicals, grain and cement.

Many US ports served by cruise ships. Popular destinations include the Caribbean, the Mexican Riviera, Hawaii and the Inside Passage to Alaska. Automobile ferries operate in many locations where bridges are impractical and in congested metropolitan areas, including New York City and San Francisco Bay.

Waterways

The United States has 25,482 miles (41,009 km) of navigable inland channels (rivers and canals), exclusive of the Great Lakes. Out of this 12,006 miles (19,322 km) is used in commerce. About 24,150 km (15,000 mi) of the Mississippi River system are presently navigable, although not all is used for commerce. The Saint Lawrence Seaway of 2,342 miles (3,769 km), including the Saint Lawrence River of 1,900 miles (3,100 km), is shared with Canada.

Ports and Harbours

United States ports and harbours include:

- Anchorage, Alaska
- Baltimore, Maryland
- Boston, Massachusetts
- Charleston, South Carolina
- Chicago, Illinois
- Duluth, Minnesota
- Erie, Pennsylvania
- Hampton Roads, Virginia

- Honolulu, Hawaii
- Houston, Texas
- Jacksonville, Florida
- Long Beach, California
- Los Angeles, California
- Miami, Florida
- Mobile, Alabama
- New Orleans, Louisiana
- Newark-New York City
- Philadelphia, Pennsylvania
- Pittsburgh, Pennsylvania
- Port Canaveral, Florida
- Portland, Oregon
- Providence, Rhode Island
- Oakland, California
- San Diego, California
- Savannah, Georgia
- Seattle, Washington
- Tacoma, Washington
- Tampa, Florida
- Toledo, Ohio
- Valdez, Alaska.

8

Merchant Marine

Most US exports and imports are on foreign ships. The Jones Act bars foreign ships from US internal trade, thus creating this "Jones Act fleet". Deck officers and ship's engineers of US flagged ships are usually trained at one of the established maritime academies. United States Merchant Marine

The United States Merchant Marine refers to the fleet of U.S. Civilian-owned merchant ships, operated by either the government or the private sector, that are engaged in commerce or transportation of goods and services in and out of the navigable waters of the United States. The Merchant Marine is responsible for transporting cargo and passengers during peace time. In time of war, the Merchant Marine is an auxiliary to the Navy, and can be called upon to deliver troops and supplies for the military.

Overview

The merchant marine is a civilian auxiliary of the U.S. Navy, but not a uniformed service, except in times of war when, in accordance with the Merchant Marine Act of 1936, mariners are considered military personnel. In 1988, President Ronald Reagan signed a bill into law granting veteran status to merchant mariners who served in war. Prior to such legislation, they were considered civilians and did not receive any of the veteran's benefits given to members of the U.S. armed forces.

As of 2006, the United States merchant fleet numbered 465 ships and approximately 69,000 members. Seven hundred ships owned by American interests but registered, or flagged, in other countries are not included in this number.

The federal government maintains fleets of merchant ships via organizations such as Military Sealift Command and the National Defence Reserve Fleet. In 2004, the federal government employed approximately 5% of all American water transportation workers.

In the 19th and 20th centuries, a number of laws were enacted that fundamentally changed the course of American merchant shipping. These laws put an end to practices such as flogging and shanghaiing, and increased shipboard safety and the standard of living. The United States Merchant Marine is also governed by several international conventions to promote safety and prevent pollution.

Background

Merchant mariners move cargo and passengers between nations and within the United States, operate and maintain deep-sea merchant ships, tugboats, towboats, ferries, dredges, excursion vessels, and other waterborne craft on the oceans, the Great Lakes, rivers, canals, harbours, and other waterways.

Captains, mates, and pilots supervise ship operations on domestic waterways and the high seas. A captain is in overall command of a vessel, and supervises the work of all other officers and crew. The captain orders the ship's course and speed, maneuvers to avoid hazards, and continuously monitors the ship's position. Captains oversee crew members who steer the vessel, determine its location, operate engines, communicate with other vessels, perform maintenance, handle lines, and operate the ship's equipment. Captains and their department heads ensure that proper procedures and safety practices are followed, ensure that machinery is in good working order, and oversee the loading and discharging of cargo and passengers. They also maintain logs and other records tracking the ships' movements, efforts at controlling pollution, and cargo and passengers carried.

The mates direct a ship's routine operation for the captain during the shifts, which are called watches. Mates stand watch for specified periods, usually 4 hours on and 8 hours off. When more than one mate is necessary aboard a ship, they typically are designated chief mate or first mate, second mate, third mate, and so forth. Mates also supervise the ship's crew. They monitor cargo loading and unloading to ensure proper stowage, and supervise crew members engaged in maintenance and the vessel's upkeep.

Pilots guide ships in and out of confined waterways, such as harbours, where a familiarity with local conditions is of prime importance. Harbour pilots are generally independent contractors who accompany vessels while they enter or leave port, and may pilot many ships in a single day.

Ship's engineers operate, maintain, and repair propulsion engines, boilers, generators, pumps, and other machinery. Merchant marine vessels usually have four engineering officers: A chief engineer and a first, second, and third assistant engineer. On many ships, Assistant Engineers stand periodic watches, overseeing the safe operation of engines and machinery. However, most modern ships sailing today utilize Unmanned Machinery Space (UMS) automation technology, and Assistant Engineers are Dayworkers. At night and during meals and breaks the engine room is unmanned and machinery alarms are answered by the Duty Engineer.

Deck officers and ship's engineers are usually trained at maritime academies. However, women were barred from entry to U.S. Maritime academies until 1974, when the State University of New York Maritime College and the California Maritime Academy admitted women as cadets. It is becoming increasingly difficult for unlicensed mariners to earn a merchant marine license due to increased requirements for formal training. To do so, a mariner must have sufficient sea time in a qualified rating and complete specified testing and training, such as that required by the International Convention on Standards of Training, Certification and Watchkeeping for Seafarers (STCW).

Able seamen and ordinary seamen operate the vessel and its deck equipment under the officers' supervision and keep their assigned areas in good condition. They stand watch, looking out for other vessels and obstructions in the ship's path, as well as for navigational aids such as buoys and lighthouses. They also steer the ship, measure water depth in shallow water, and maintain and operate deck equipment such as lifeboats, anchors, and cargo-handling gear. On tankers, mariners designated as pumpmen hook up hoses, operate pumps, and clean tanks. When arriving at or leaving a dock, they handle the mooring lines. Seamen also perform routine maintenance chores, such as repairing lines, chipping rust, and painting and cleaning decks. On larger vessels, a boatswain, or head seaman will supervise the work.

Marine oilers and more experienced qualified members of the engine department, or QMEDs, maintain the vessel in proper running order in the engine spaces below decks, under the direction of the ship's engineering officers. These workers lubricate gears, shafts, bearings, and other moving parts of engines and motors; read pressure and temperature gauges; record data; and sometimes assist with repairs and adjust machinery. Wipers are the entry-level workers in the engine room, holding a position similar to that of ordinary seamen of the deck crew. They clean and paint the engine room and its equipment and assist the others in maintenance and repair work. With more experience they become oilers and firemen.

A typical deep-sea merchant ship has a captain, three mates, a chief engineer and three assistant engineers, plus six or more unlicensed seamen, such as able seamen, oilers, QMEDs, and cooks or food handlers. Other unlicensed positions on a large ship may include electricians and machinery mechanics.

History

The history of ships and shipping in North America goes back at least as far as when Leif Erikson established a short-lived settlement called Vinland in present day Newfoundland. An actual shipping industry gradually came into being as colonies grew and trade with Europe increased. As early as the 16th century, Europeans were shipping horses, cattle and hogs to the Americas.

Spanish colonies began to form as early as 1565 in places like St. Augustine, Florida, and later in Santa Fe, New Mexico,

San Antonio, Tucson, San Diego, Los Angeles and San Francisco. English colonies like Jamestown began to form as early as 1607. The connection between the American colonies and Europe, with shipping as its only conduit, would continue to grow unhindered for almost two hundred years.

The first wartime role of an identifiable United States merchant marine first took place on June 12, 1775 in and around Machias, Maine. A group of citizens, hearing the news from Concord and Lexington, captured the British schooner HMS Margaretta. The citizens, in need of critical supplies, were given an ultimatum: either load the ships with lumber to build British barracks in Boston, or go hungry. They chose to fight.

Word of this revolt reached Boston, where the Continental Congress and the various colonies issued Letters of Marque to privateers. The privateers interrupted the British supply chain all along the eastern seaboard of the United States and across the Atlantic Ocean. These actions by the privateers predates both the United States Coast Guard and the United States Navy, which were formed in 1790 and 1775, respectively.

The merchant marine was active in subsequent wars, from the Confederate commerce raiders of the American Civil War, to the First and Second Battle of the Atlantic in World War I and World War II. 3.1 million tons of merchant ships were lost in World War II, mariners dying at a rate of 1 in 24, which was the highest rate of casualties of any service. All told, 733 American cargo ships were lost and 8,651 of the 215,000 who served perished on troubled waters and off enemy shores.

Merchant shipping also played its role in the wars in Vietnam and Korea. From just six ships under charter when the Korean war began, this total peaked at 255. In September 1950, when the U.S. Marine Corps went ashore at Inchon, 13 USNS cargo ships, 26 chartered American, and 34 Japanese-manned merchant ships, under the operational control of Military Sea Transportation Service participated in the invasion.

During the Vietnam War, ships crewed by civilian seamen carried 95% of the supplies used by the American armed forces. Many of these ships sailed into combat zones under fire. In fact, the SS Mayaguez incident involved the capture of mariners from the American merchant ship SS Mayaguez.

During the first Gulf War, the merchant ships of the Military Sealift Command (MSC) delivered more than 11 million metric tons of vehicles, helicopters, ammunition, fuel and other supplies and equipment. At one point during the war, more than 230 government-owned and chartered ships were involved in the sealift.

Government-owned merchant vessels from the National Defence Reserve Fleet (NDRF) have supported emergency shipping requirements in seven wars and crises. During the Korean War, 540 vessels were activated to support military forces. A worldwide tonnage shortfall from 1951 to 1953 required over 600 ship activations to lift coal to Northern Europe and grain to India. From 1955 through 1964, another 600 ships were used to store grain for the Department of Agriculture. Another tonnage shortfall following the Suez Canal closing in 1956 caused 223 cargo ship and 29 tanker activations from the NDRF. During the Berlin crisis of 1961, 18 vessels were activated, which remained in service until 1970. The Vietnam conflict required the activation of 172 vessels.

Since 1977, the Ready Reserve Fleet has taken over the brunt of the work previously handled by the National Defence Reserve Fleet. The RRF made a major contribution to the success of Operation Desert Shield/Operation Desert Storm from August 1990 through June 1992, when 79 vessels were activated to meet military sealift requirements by carrying 25% of the unit equipment and 45% of the ammunition needed.

Two RRF tankers, two RO/RO ships and a troop transport ship were needed in Somalia for Operation Restore Hope in 1993 and 1994. During the Haitian crisis in 1994, 15 ships were activated for Operation Uphold Democracy operations. In 1995 and 1996, four RO/RO ships were used to deliver military cargo as part of U.S. and U.K. support to NATO peace-keeping missions.

Four RRF ships were activated to provide humanitarian assistance for Central America following Hurricane Mitch in 1998. Three RRF ships currently support the Afloat

Prepositioning Force with two specialized tankers and one dry cargo vessel capable of underway replenishment for the Navy's Combat Logistics Force.

In 2003, 40 RRF ships were used in support of Operation Enduring Freedom and Operation Iraqi Freedom. This RRF contribution was significant and included sealifting equipment and supplies into the theatre of combat operations, which included combat support equipment for the Army, Navy Combat Logistics Force, and USMC Aviation Support equipment. By the beginning of May 2005, RRF cumulative support included 85 ship activations that logged almost 12,000 ship operating days, moving almost 25% of the equipment needed to support the U.S. Armed Forces liberation of Iraq.

MSC is also involved in the current Iraq War, having delivered 61 million square feet (5.7 km^2) of cargo and 1.1 billion US gallons (4,200,000 m^3) of fuel by the end of the first year alone. Merchant mariners are being recognized for their contributions in Iraq. For example, in late 2003, Vice Adm. David Brewer III, commander of Military Sealift Command, awarded the officers and crewmembers of the MV Capt Steven L. Bennett the Merchant Marine Expeditionary Medal.

The RRF was called upon to provide humanitarian assistance to gulf coast areas following Hurricane Katrina and Hurricane Rita landfalls in September 2005. The Federal Emergency Management Agency requested a total of eight vessels to support relief efforts. Messing and berthing was provided for refinery workers, oils spill response teams, longshoremen. One of the vessels provided electrical power.

• Today's Merchant Fleet

The Commercial Fleet

As of 2006, the United States merchant fleet had 465 privately-owned ships of 1,000 gross register tons or over. Two hundred ninety-one (291) of these were dry cargo ships, 97 were tankers, and 77 passenger ships. Of those American-flagged ships, 51 were foreign owned. Seven hundred American-owned ships are flagged in other nations. 2005 statistics from the United States Maritime Administration focus on the larger segment of the fleet: ships of 10,000 metric tons deadweight (DWT) and over. 245 privately owned American-flagged ships are of this size, and 153 of those meet the Jones Act criteria.

U.S. sealift capability viewed over time shows a steep drop in the number of ships in the merchant marine fleet. Observers point to the World War II era as the peak for the U.S. fleet. During the post-war year of 1950, for example, U.S. carriers represented about 43 percent of the world's shipping trade. By 1995, the American market share had plunged to 4 percent, according to a 1997 report by the U.S. Congressional Budget Office (CBO). CBO further notes in the report that "the number of U.S.-flag vessels has dropped precipitously—from more than 2,000 in the 1940s and 850 in 1970 to about 320 in 1996."

A diminishing U.S. fleet comes in the face of surge in international sea trade. For instance, worldwide demand for natural gas and the subsequent spike in related international trade presents a job growth opportunity for today's U.S. mariners aboard liquefied natural gas (LNG) tankers. A 2007 agreement signed by the United States Maritime Administration (MARAD) sets uniform LNG training standards at U.S. maritime training facilities. Uniform training standards will help U.S. mariners compete for jobs aboard LNG tankers, estimated to number more than 370 worldwide at the close of 2007, according to MARAD.

However, despite projection of an eightfold increase in U.S. imported LNG by 2025, the worldwide LNG fleet does not include a single U.S. flagged vessel. Moreover, only five U.S. deepwater LNG ports were operational in 2007, although permits have been issued for four additional ports, according to MARAD.

Further limiting potential job growth in the U.S. fleet is the fact that ranks of qualified seamen to serve on ships continue to shrink. Recruitment efforts to attract younger mariners to replace retiring crews have failed to stem the shortage. MARAD describes the gap between sealift crewing needs and available unlicensed personnel as "reaching critical proportions, and the long term outlook for sufficient personnel is also of serious concern." Seagoing jobs of the future for U.S. mariners may not necessarily be on U.S.-flagged ships. American-trained mariners are being sought after by international companies to operate foreign-flagged vessels, according to Julie A. Nelson, deputy maritime administrator of the U.S. Department of Commerce. For example, Shell International and Shipping Company Ltd. has announced that it will be recruiting U.S. seafarers to crew its growing fleet of tankers. Further signs of the globalization of the mariner profession is evidenced by an agreement signed in 2007 between Overseas Shipholding Group and the Maritime Administration that will allow American maritime academy cadets to train aboard OSG's international flag vessels.

The Federal Fleet

Military Sealift Command (MSC) is an arm of the Navy that serves the entire Department of Defence as the ocean carrier of materiel during peacetime and war. It transports equipment, fuel, ammunition, and other goods essential to the smooth function of United States armed forces worldwide. Up to 95% of all supplies needed to sustain the U.S. military can be moved by Military Sealift Command. MSC operates approximately 120 ships with 100 more in reserve. All ships are manned by civil service or contract merchant mariners, estimated to number more than 8,000. MSC tankers and freighters have a long history of also serving as re-supply vessels in support of civilian research at McMurdo Station, Antarctica, and at other polar operations, including Greenland.

The National Defence Reserve Fleet (NDRF) acts as a reserve of cargo ships for national emergencies and defence. Consisting of 2,277 ships at its peak in 1950, the NDRF fleet now numbers only 251 ships.

NDRF vessels are now staged at the James River, Beaumont and Suisun Bay fleet sites and other designated locations. A Ready Reserve Force component of NDRF was established in 1976 to provide rapid deployment of military equipment. This force currently has 58 vessels, down from a peak of 102 in 1994.

In 2004, the Federal government employed approximately 5% of all water transportation workers, most of whom worked on Military Sealift Command supply ships.

Important Laws

A few laws have shaped the development of the U.S. merchant marine. Chief among them are the "Seamen's Act of 1915," the "Merchant Marine Act of 1920" (commonly referred to as the "Jones Act"), and the "Merchant Marine Act of 1936."

The Seamen's Act of 1915

The Seaman's Act significantly improved working conditions for American seamen. The brainchild of International Seamen's Union president Andrew Furuseth, the Act was sponsored in the Senate by Robert Marion La Follette and received significant support from Secretary of Labour William B. Wilson.

Among other things, the Act:

1. abolished the practice of imprisonment for seamen who deserted their ship
2. reduced the penalties for disobedience
3. regulated a seaman's working hours both at sea and in port
4. established a minimum quality for ship's food
5. regulated the payment of seamen's wages
6. required specific levels of safety, particularly the provision of lifeboats
7. required a minimum percentage of the seamen aboard a vessel to be qualified Able Seamen
8. required a minimum of 75% of the seamen aboard a vessel to understand the language spoken by the officers.

The Act's passage was attributed to union lobbying, increased tensions immediately before World War I, and raised public consciousness of safety at sea due to the sinking of the RMS Titanic three years prior.

The Jones Act

The "Merchant Marine Act of 1920," often called The "Jones Act," requires U.S.-flagged vessels be built in the United States, owned by U.S. citizens, and documented under the laws of the United States. It also requires that all officers and 75% of the crew must be U.S. citizens. Vessels satisfying these requirements comprise the "Jones Act Fleet," and only these vessels may engage in "cabotage", or carrying passengers or cargo between two U.S. ports.

Another important aspect of the Act is that it allows injured sailors to obtain damages from their employers for the negligence of the shipowner, the captain, or fellow members of the crew.

The Merchant Marine Act

The Merchant Marine Act of 1936 was enacted "to further the development and maintenance of an adequate and well-balanced American merchant marine, to promote the commerce of the United States, to aid in the national defence, to repeal certain former legislation, and for other purposes."

Specifically, the Act established the United States Maritime Commission and required a United States Merchant Marine that consists of U.S.-built, U.S.-flagged, U.S.-crewed and U.S.-owned vessels capable of carrying all domestic and a substantial portion of foreign water-borne commerce which can serve as a naval auxiliary in time of war or national emergency.

The act also established federal subsidies for the construction and operation of merchant ships. Two years after the Act was passed, the U.S. Merchant Marine Cadet Corps, the forerunner to the United States Merchant Marine Academy, was established.

International Regulations

Federal law requires the merchant marine to adhere to a number of international conventions. The International Maritime Organization has been either the source or a conduit for a number of these regulations.

The principal International Conventions are:

- SOLAS 74: International Convention for the Safety of Life at Sea.
- MARPOL 73/78: International Convention for the Prevention of Pollution From Ships, 1973 as modified by the Protocol of 1978.

- ICLL 66: International Convention on Load Lines, as revised in 1966
- 72 COLREGS: International Regulations for Preventing Collisions at Sea.
- STCW 95: International Convention on Standards of Training, Certification and Watchkeeping for Seafarers (STCW).
- SAR 79: International Convention on Maritime Search and Rescue.

Military

The federal military has a dedicated system of bases with runways, aircraft, watercraft, conventional cars and trucks, and armored and special-purpose vehicles. During times of war, it may commandeer private infrastructure and vehicles as authorized by Congress and the President.

Pipeline Statistics:

- Petroleum products: 224,620 km
- Natural gas: 548,665 km (2006)

Funding

Federal, state, and local tax revenues support upkeep of most roads, which are generally free to drivers. There are also some toll roads and toll bridges. Most other forms of transportation charge a fee for use as they are not given much, if any, tax support by Congress.

Government funding of transportation exists at many levels. Federal funding for highway, rail, bus, water, air, and other forms of transportation is allocated by Congress for several years at a time. The current authorization bill is the Safe, Accountable, Flexible, Efficient Transportation Equity Act: A Legacy for Users (SAFETEA-LU), which runs from 2005 to 2009. A Congressionally chartered committee is considering future funding issues.

Though earmarks are often made for specific projects, the allocation of most federal dollars is controlled by metropolitan planning organizations (MPOs) and state governments. Usually "matching" funds are required from local sources. All projects have a sponsoring agency that will receive the funding from the various federal and local sources, and be responsible for implementing the project directly or through contracts.

Large projects require a Major Investment Study and a both a Draft and a Final Environmental Impact Review. A patchwork of federal laws and accounts govern the allocation of federal transportation dollars, most of which is reserved for capital projects, not operating expenses. Some roads are federally designated as part of the National Highway System and get preferential funding as a result, but there are few federally maintained roads outside of Washington D.C. and national parks.

State governments are sovereign entities which use their powers of taxation both to match federal grants, and provide for local transportation needs. Different states have different systems for dividing responsibility for funding and maintaining road and transit networks between the state department of transportation, counties, municipalities, and other entities. Typically cities or counties are responsible for local roads, financed with block grants and local property taxes, and the state is responsible for major roads that receive state and federal designations. Many mass transit agencies are quasi-independent and subsidized branches of a state, county, or city government.

Economic Impact

According to the U.S. Department of Transportation (DOT): "Transportation's vital importance to the U.S. economy is underscored by the fact that more than $1 out of every $10 produced in the U.S. gross domestic product is related to transportation activity.

This includes all aspects of transportation, including the movement of goods and the purchase of all transportation-related products and services as well as the movement of people." Employment in the transportation and material moving industry accounted for 7.4% of all employment, and was the 5th largest employment group in the United States.

Environmental Impacts

Two-thirds of U.S. oil consumption is due to the transportation sector. The "Energy Independence and Security Act of 2007" has a significant impact on U.S. Energy Policy. The US — an important export country for food stocks — will convert 18% of its grain output to ethanol in 2008. Across the US, 25% of the whole corn crop went to ethanol in 2007. The percentage of corn going to biofuel is expected to go up. In 2006, U.S. Senators introduced the BioFuels Security Act.

Everyday personal transportation in America is dominated by the automobile. As of 2003, there were 759 automobiles per 1,000 Americans, compared to 472 per 1,000 inhabitants of the European Union the following year. About 40% of personal vehicles are vans, SUVs, or light trucks. The average American adult (accounting for all drivers and nondrivers) spends 55 minutes driving every day, traveling 29 miles (47 km).

The civil airline industry is entirely privately owned, while most major airports are publicly owned. The four largest airlines in the world by passengers carried are American; Southwest Airlines is number one. Of the world's thirty busiest passenger airports, sixteen are in the United States. It is also home to the busiest airport in the world, Hartsfield-Jackson Atlanta International Airport. While transport of goods by rail is extensive, relatively few people use rail to travel, within or between cities. Mass transit accounts for 9% of total U.S. work trips, compared to 38.8% in Europe. Bicycle usage is minimal, well below European levels.

Energy

The United States energy market is 29,000 terawatt hours per year. Energy consumption per capita is 7.8 tons of oil equivalent per year, compared to Germany's 4.2 tons and Canada's 8.3 tons. In 2005, 40% of this energy came from petroleum, 23% from coal, and 22% from natural gas. The remainder was supplied by nuclear power and renewable energy sources. The United States is the world's largest consumer of petroleum. For decades, nuclear power has played a limited role relative to many other developed countries, in part due to public perception in the wake of the 1979 Three Mile Island accident. In 2007, several applications for new nuclear plants were filed.

Demographics

The United States population is projected by the U.S. Census Bureau to be 309,612,000, including an estimated 11.2 million illegal immigrants. The third most populous nation in the world, after China and India, the United States is the only industrialized nation in which large population increases are projected. With a birth rate of 13.82 per 1,000, 30% below the world average, its population growth rate is 0.98%, significantly higher than those of Western Europe, Japan, and South Korea.

In fiscal year 2009, 1.1 million immigrants were granted legal residence. Mexico has been the leading source of new residents for over two decades; since 1998, China, India, and the Philippines have been in the top four sending countries every year.

The United States has a very diverse population—thirty-one ancestry groups have more than a million members. White Americans are the largest racial group; German Americans, Irish Americans, and English Americans constitute three of the country's four largest ancestry groups. African Americans are the nation's largest racial minority and third largest ancestry group. Asian Americans are the country's second largest racial minority; the two largest Asian American ethnic groups are Chinese American and Filipino American. In 2008, the U.S. population included an estimated 4.9 million people with some American Indian or Alaskan native ancestry (3.1 million exclusively of such ancestry) and 1.1 million with some native Hawaiian or Pacific island ancestry (0.6 million exclusively).

The population growth of Hispanic and Latino Americans (the terms are officially interchangeable) is a major demographic trend. The 46.9 million Americans of Hispanic descent are identified as sharing a distinct "ethnicity" by the Census Bureau; 64% of Hispanic Americans are of Mexican descent. Between 2000 and 2008, the country's Hispanic population increased 32% while the non-Hispanic population rose just 4.3%. Much of this growth is from immigration; as of 2007, 12.6% of the U.S. population was foreign-born, with 54% of that figure born in Latin America. Fertility is also a factor; the average Hispanic woman gives birth to three children in her lifetime. The comparable fertility rate is 2.2 for non-Hispanic black women and 1.8 for non-Hispanic white women (below the replacement rate of 2.1). Minorities (as defined by the Census Bureau, all those beside non-Hispanic, non-multiracial whites) constitute 34% of the population; they are projected to be the majority by 2042.

About 82% of Americans live in urban areas (as defined by the Census Bureau, such areas include the suburbs); about half of those reside in cities with populations over 50,000. In 2008, 273 incorporated places had populations over 100,000, nine cities had more than 1 million residents, and four global cities had over 2 million (New York City, Los Angeles, Chicago, and Houston). There are fifty-two metropolitan areas with populations greater than 1 million. Of the fifty fastest-growing metro areas, forty-seven are in the West or South. The metro areas of Dallas, Houston, Atlanta, and Phoenix all grew by more than a million people between 2000 and 2008.

Language

English is the de facto national language. Although there is no official language at the federal level, some laws—such as U.S. naturalization requirements—standardize English. In 2006, about 224 million, or 80% of the population aged five years and older, spoke only English at home. Spanish, spoken by 12% of the population at home, is the second most common language and the most widely taught second language. Some Americans advocate making English the country's official language, as it is in at least twenty-eight states. Both Hawaiian and English are official languages in Hawaii by state law. While neither has an official language, New Mexico has laws providing for the use of both English and Spanish, as Louisiana does for English and French. Other states, such as California, mandate the publication of Spanish versions of certain government documents including court forms. Several insular territories grant official recognition to their native languages, along with English: Samoan and Chamorro are recognized by American Samoa and Guam, respectively; Carolinian and

Chamorro are recognized by the Northern Mariana Islands; Spanish is an official language of Puerto Rico.

Religion

The United States is officially a secular nation; the First Amendment of the U.S. Constitution guarantees the free exercise of religion and forbids the establishment of any religious governance. In a 2002 study, 59% of Americans said that religion played a "very important role in their lives," a far higher figure than that of any other wealthy nation. According to a 2007 survey, 78.4% of adults identified themselves as Christian, down from 86.4% in 1990. Protestant denominations accounted for 51.3%, while Roman Catholicism, at 23.9%, was the largest individual denomination. The study categorizes white evangelicals, 26.3% of the population, as the country's largest religious cohort; another study estimates evangelicals of all races at 30–35%. The total reporting non-Christian religions in 2007 was 4.7%, up from 3.3% in 1990. The leading non-Christian faiths were Judaism (1.7%), Buddhism (0.7%), Islam (0.6%), Hinduism (0.4%), and Unitarian Universalism (0.3%). The survey also reported that 16.1% of Americans described themselves as agnostic, atheist, or simply having no religion, up from 8.2% in 1990.

Education

American public education is operated by state and local governments, regulated by the United States Department of Education through restrictions on federal grants. Children are required in most states to attend school from the age of six or seven (generally, kindergarten or first grade) until they turn eighteen (generally bringing them through twelfth grade, the end of high school); some states allow students to leave school at sixteen or seventeen. About 12% of children are enrolled in parochial or nonsectarian private schools. Just over 2% of children are homeschooled. The United States has many competitive private and public institutions of higher education, as well as local community colleges with open admission policies. Of Americans twenty-five and older, 84.6% graduated from high school, 52.6% attended some college, 27.2% earned a bachelor's degree, and 9.6% earned graduate degrees. The basic literacy rate is approximately 99%. The United Nations assigns the United States an Education Index of 0.97, tying it for 12th in the world.

Health

The United States life expectancy of 77.8 years at birth is a year shorter than the overall figure in Western Europe, and three to four years lower than that of Norway, Switzerland, and Canada. Over the past two decades, the country's rank in life expectancy has dropped from 11th to 42nd in the world. The infant mortality rate of 6.37 per thousand likewise places the United States 42nd out of 221 countries, behind all of Western Europe. Approximately one-third of the adult population is obese and an additional third is overweight; the obesity rate, the highest in the industrialized world, has more than doubled in the last quarter-century. Obesity-related type 2 diabetes is considered epidemic by health care professionals. The U.S. adolescent pregnancy rate, 79.8 per 1,000 women, is nearly four times that of France and five times that of Germany. Abortion, legal on demand, is highly controversial. Many states ban public funding of the procedure and restrict late-term abortions, require parental notification for minors, and mandate a waiting period. While the abortion rate is falling, the abortion ratio of 241 per 1,000 live births and abortion rate of 15 per 1,000 women aged 15–44 remain higher than those of most Western nations.

The U.S. health care system far outspends any other nation's, measured in both per capita spending and percentage of GDP. The World Health Organization ranked the U.S. health care system in 2000 as first in responsiveness, but 37th in overall performance. The United States is a leader in medical innovation. In 2004, the nonindustrial sector spent three times as much as Europe per capita on biomedical research.

Unlike in all other developed countries, health care coverage in the United States is not universal. In 2004, private insurance paid for 36% of personal health expenditures, private out-of-pocket payments covered 15%, and federal, state, and local governments paid for 44%. In 2005, 46.6 million Americans, 15.9% of the population, were uninsured, 5.4 million more than in 2001. The main cause of this rise is the drop in the number of Americans with employer-sponsored health insurance. The subject of uninsured and underinsured Americans is a major political issue. A 2009 study estimated that lack of insurance is associated with nearly 45,000 deaths a year. In 2006, Massachusetts became the first state to mandate universal health insurance. Federal legislation passed in early 2010 will create a near-universal health insurance system around the country by 2014.

Crime and Law Enforcement

Among developed nations, the United States has above-average levels of violent crime and particularly high levels of gun violence and homicide. In 2007, there were 5.6 murders per 100,000 persons, three times the rate in neighbouring Canada. The U.S. homicide rate, which decreased by 42% between 1991 and 1999, has been roughly steady since. Gun ownership rights are the subject of contentious political debate.

The United States has the highest documented incarceration rate and total prison population in the world. At the start of 2008, more than 2.3 million people were incarcerated, more than one in every 100 adults. The current rate is about seven times the 1980 figure. African American males are jailed at about six times the rate of white males and three times the rate of Hispanic males. In 2006, the U.S. incarceration rate was over three times the figure in Poland, the Organisation for Economic Co-operation and Development (OECD) country with the next highest rate. The country's high rate of incarceration is largely due to sentencing and drug policies.

Though it has been abolished in most Western nations, capital punishment is sanctioned in the United States for certain federal and military crimes, and in thirty-six states. Since 1976, when the U.S. Supreme Court reinstated the death penalty after a four-year moratorium, there have been more than 1,000 executions. In 2006, the country had the sixth highest number of executions in the world, following China, Iran, Pakistan, Iraq, and Sudan. In 2007, New Jersey became the first state to legislatively abolish the death penalty since the 1976 Supreme Court decision, followed by New Mexico in 2009.

Culture

The United States is a multicultural nation, home to a wide variety of ethnic groups, traditions, and values. Aside from the now small Native American and Native Hawaiian populations, nearly all Americans or their ancestors immigrated within the past five centuries. The culture held in common by most Americans—mainstream American culture—is a Western culture largely derived from the traditions of European immigrants with influences from many other sources, such as traditions brought by slaves from Africa. More recent immigration from Asia and especially Latin America has added to a cultural mix that has been described as both a homogenizing melting pot and a heterogeneous salad bowl in which immigrants and their descendants retain distinctive cultural characteristics.

According to Geert Hofstede's cultural dimensions analysis, the United States has the highest individualism score of any country studied. While the mainstream culture holds that the United States is a classless society, scholars identify significant differences between the country's social classes, affecting socialization, language, and values. The American middle and professional class has initiated many contemporary social trends such as modern feminism, environmentalism, and multiculturalism. Americans' self-images, social viewpoints, and cultural expectations are associated with their occupations to an unusually close degree. While Americans tend greatly to value socioeconomic achievement, being ordinary or average is generally seen as a positive attribute. Though the American Dream, or the perception that Americans enjoy high social mobility, plays a key role in attracting immigrants, some analysts find that the United States has less social mobility than Western Europe and Canada.

Women now mostly work outside the home and receive a majority of bachelor's degrees. In 2007, 58% of Americans age 18 and over were married, 6% were widowed, 10% were divorced, and 25% had never been married. Same-sex marriage is contentious. Some states permit civil unions in lieu of marriage. Since 2003, several states have permitted gay marriage as the result of judicial or legislative action, while voters in more than a dozen states have barred the practice via referendum.

Social Class in the United States

Social class certainly exists in the United States, though it remains a controversial concept with many competing definitions. Many Americans believe in a simple three-class model that includes the "rich", the "middle class", and the "poor." More complex models that have been proposed describe as many as a dozen class levels; while still others deny the very existence, in the strict sense, of "social class" in American society. Most definitions of class structure group people according to wealth, income, education, type of occupation, and membership in a specific subculture or social network.

Dennis Gilbert, William Thompson, Joseph Hickey, and James Henslin have proposed class systems with six distinct social classes. These class models feature an upper or capitalist class consisting of the rich and powerful, an upper middle class consisting of highly-educated and wealthy professionals, a middle class consisting of college-educated individuals employed in white collar industries, a lower middle class, a working class constituted by clerical and blue collar workers whose work is highly routinized, and a lower class divided between the working poor and the unemployed underclass.

A Classless Society?

Some academics consider American society sociologically and economically fragmented in such a manner that no clear class distinctions can be made. This means that there are no pronounced breaks in socioeconomic strata, which makes class division highly subjective and disputable. Others, such as sociologist Dennis Gilbert, dispute the concept of a well-mixed society, and claim that distinct social networks can be identified for each class. W. Lloyd Warner also asserts the existence of class markers:

"We are proud of those facts of American life that fit the pattern we are taught but somehow we are often ashamed of those equally important social facts which demonstrate the presence of social class. Consequently, we tend to deny them, or worse, denounce them and by doing so we tend to deny their existence and magically make them disappear from consciousness."

—*W. Lloyd Warner*, What Social Class Is In America

Warner asserts that social class is as old as civilization itself and has been present in nearly every society from before the Roman Empire, through medieval times, and to the modern-day United States. He believes that complex societies such as the United States need an equally complex social hierarchy.

Markers

Some definitions of class look only at numerical measures such as wealth or income. Others take into account qualitative factors, such as education, culture, and social status. There is no consensus on which of these variables is essential and which are merely common correlates. It is also disputed whether sharp lines can be drawn; one point of view in the debate:

> *"A stratified society is one marked by inequality, by differences among people that are regarded as being higher or lower...it is logically possible for a society to be stratified in a continuous gradation between high and low without any sharp lines...in reality...there is only a limited number of types of occupations... People in similar positions...grow similar in their thinking and lifestyle...they form a pattern, and this pattern creates social class."*

—Dennis Gilbert, The American Class Structure, 1998

Social Status

> *"It is impossible to understand people's behavior...without the concept of social stratification, because class position has a pervasive influence on almost everything...the clothes we wear...the television shows we watch...the colors we paint our homes in and the names we give our pets... Our position in the social hierarchy affects our health, happiness, and even how long we will live."*

—William Thompson, Joseph Hickey, Society in Focus, 2005

Social class is sometimes presented as a description of how the society has distributed its members among positions of varying importance, influence, and prestige. In these models, certain occupations are considered to be desirable and influential, while others are considered to be menial, repetitive, and unpleasant. (In some cases, non-occupational roles such as a parent or volunteer mentor, are also considered.) Higher-class jobs require more skill and education.

Some sociologists consider the higher income and prestige of higher-class jobs to simply be incentives to obtain the skills necessary to perform important work. This is an important mechanism in the economic theory of capitalism, and is compatible with the notion that class is mutable and determined by a combination of choices and opportunities.

In other cases, class or status is inherited. For example, being the son or daughter of a wealthy individual, may carry a higher status and different cultural connotations than being a member of nouveau riche ("new money"). Those taking the functionalist approach to sociology and economics view social classes as components essential for the survival of complex societies such as American society.

Income

Income in the United States is most commonly measured by U.S. Census Bureau in terms of either household or individual and remains one of the most prominent indicators of class status. As 42% of all households, 76% of those in the top quintiles, had two income earners the discrepancy between household and personal income is quite considerable. In 2005 the top 15% of income earners made $62,500 or more, while 18% of households had incomes over $100,000. Personal income is largely the result of scarcity. As individuals who hold higher status positions tend to possess rare skills or assume positions society deems very essential, have higher incomes. Overall the median household income was $46,326 in 2005 while the median personal income (including only those above the age of 25) was $32,140.

Per capita household income, the income a household is able to allocate to each member of the household is also an important variable in determining a given household's standard of living. A high household income may be offset by a large household size; thus, resulting in a low per capita household income. In 2005, the median household income per capita was $24,672.

> *"It should be stressed...that a position does not bring power and prestige because it draws a high income. Rather, it draws a high income because it is functionally important and the available personnel are for one reason or another scarce. It is therefore superficial and erroneous to regard high income as the cause of a man's power and prestige, just as it is erroneous to think that a man's fever is the cause of his disease... The economic source of power and prestige is not income primarily, but the ownership of capital goods (including patents, good will, and professional reputation). Such ownership should be distinguished from the possession of consumers' goods, which is an index rather than a cause of social standing."*
>
> —Kingsley Davis and Wilbert E. Moore, Principles of Stratification

As mentioned above, income is one of the most prominent features of social class; it is not one of its causes. In other words, income does not determine the status of an individual or household but rather reflects on that status. Some say that income and prestige are the incentives provided by society in order to fill needed positions with the most qualified and motivated personnel possible. However such a conclusion is contrary to the widespread observation that upper management at many companies is less knowledgeable and competent than many of their underlings, and that their positions are largely the result of fortuitous birth situations and networking. This situation is lampooned in the popular social commentary comic strip "Dilbert".

The New York Times has used income quintiles to define class. It has assigned the quintiles from lowest to highest as lower class, lower middle class, middle class, upper middle class, and upper class. These definitions equate class with income, permitting people to move from class to class as their income changes.

Dual Income Controversy

Income is one of the most commonly used attributes of a household to determine its class status. The relationship between income, which mostly arises from the scarcity of a certain skill, may however, prove to be more complex than initially perceived. While the idea is that income reflects status, household income may just be the product of two or more incomes.

In 2005, 42% of American households had two income earners. The vast majority (77%) of households in the top quaintly had two or more income earners. This means that the majority of household income in the top quaintly are the result of two income earners pooling their resources, establishing a close link between perceived affluence and the number of income earners in a given household. This raises the question of whether or not the combination of incomes results in higher social status. Of course, there is no definite answer as class is a vague sociological concept.

> *"The parade of income earners with height representing income suggest that the relationship between the distribution of income and the class structure is... blurred in the middle... we saw dual-income working class marchers looking down on single-income upper-middle class marchers. In sum, the class structure as we have defined it... does not exactly match the distribution of household income."*
>
> —Dennis Gilbert, The American Class Structure, 1998

Sociologist Dennis Gilbert states that it is possible for households to out-earn other households over higher class standing through increasing their number of income earners. He furthermore states that household size also played an essential role, as the standard of living for two persons living off one upper middle class personal income may very well be higher than that of a household with four members living off two working class personal incomes. The combination of two or more incomes, allow for households to increase their income substantially without moving higher on the occupational ladder or attaining higher educational degrees. Thus it is important to remember that the favourable economic position of households in the top two quintiles is in some cases the result of combined income, rather than demand for a single worker.

Education

Tertiary education (or "higher education") is required for many middle-class professions, depending on how the term middle class is to be defined. Tertiary education is rarely free, but the costs vary widely: tuition at elite private colleges often exceeds $200,000 for a four-year program. On the other hand, public colleges and universities typically charge much less (for state residents), and many, such as the University of California system, rival the elite private schools in reputation and quality.

Also, scholarships offered by universities and government do exist, and low-interest loans are available. Still, the average cost of education, by all accounts, is increasing. The attainment of post-secondary and graduate degrees is the perhaps most important feature of a middle and upper middle class person with the university being regarded as the most essential institution and gatekeeper of the professional middle class. Educational attainment is also directly linked to income.

In 2005, the vast majority of those with doctorate and professional degrees were among the nation's top 15% of income earners. Those with bachelor degrees had incomes considerably above the national median while the median income for those with some college education remained near the national median. According to U.S. Census Bureau, 9% of persons aged 25 or older had a graduate degree, 27.9% had a Bachelor's degree or more with 53% having attended college.

With 85% of the population having graduated high school, it becomes apparent that the average American does not have a college degree, but is likely to have attended college for some time and has graduated high school. Overall, educational attainment serves as the perhaps most essential class feature of most Americans, being directly linked to income and occupation.

Culture

Social classes feature their own sub-cultures and have therefore developed slightly different manners of socializing their offspring. Due to class mobility individuals may also assimilate to the culture of another class when ascending or descending in the social order. However, one does need to remember that all social classes in the United States, except the upper class, consist of tens of millions of people. Thus social classes form social groups so large that they feature considerable diversity within and any statement regarding a given social class' culture needs to be seen as a broad generalization.

Since 1970, sociologists such as Paula LeMasters and Melvin Kohl have set out repeatedly to research class based cultures. Class culture has been shown to have a strong influence on the mundane lives of people, affecting everything from the manner in which they raise their children, initiation and maintenance of romantic relationship to the colour in which they paint their houses. The strongest cultural differences seem to run along the professional middle class-working class divide. A recent increase in residential class segregation and the overall tendency of individual to associate mostly with those of equal standing as themselves has further strengthened class differences.

Parental views are the perhaps most essential factor in determining the socialization process which shapes new members of society. The values and standards used in child rearing are commonly closely related to the parent's occupational status. Parents from the professional class tend to raise their children to become curious independent thinkers, while working class parents raise their children to have a more communal perspective with a strong respect for authority. Middle class parents tend to emphasize internal standards and values while working class parents emphasize external values.

Sociologist Dennis Gilbert uses a list of values identified by Melvin Kohn to be typical of the professional middle and working class. Middle class parents values for their children and themselves included: "Consideration of Others, Self-Control, Curiosity, Happiness, Honesty, Tolerance of Nonconformity, Open to Innovation... Self-Direction." This contrasted with surveyed working class individuals, who reported: "Manners, Obedience... Neatness, Cleanliness, Strong Punishment of Deviant Behaviour, Stock to Old Ways, People not Trustworthy... Strict Leadership" as values for themselves and their children. There is a strong correlation between these values and the occupational activities of the respondents. The job characteristics of middle class respondents included: "Work Independently, Varied Tasks, Work with People or Data," versus working class parents of reported "Close Supervision and Repetitive Work..."

> *"Not once in a professional middle-class home did I see a young boy shake his father's hand in a well-taught manly gesture... Not once did I hear a middle-class parent scornfully-or even sympathetically-call a crying boy a sissy or in any way reprimand him for his tears... even as young as six or seven, the working class boys seemed more emotionally controlled-more like miniature men-than those in the middle-class families."*
>
> —Sociologist Lillian Rubin in Gilbert, 1998

Gender roles are also viewed differently by those in the higher and lower social classes. Middle class individuals, who were more open towards "nonconformity" and emphasized individual self-direction as well as critical thinking, were also less stringent in their application of gender roles. Working class individuals, on the other hand, emphasized gender roles. While working class people have more and more assimilated to middle class culture regarding their view and application of gender roles, differences remain. Professional class people are more likely to have an egalitarian distribution of work in their household with both spouses being equals in heterosexual marriages. According to Dennis Gilbert, "College life, generally a prologue to upper-middle class careers, delays marriage and encourages informal, relatively egalitarian association between men and women."

Upper Class

This term is applied to a wide array of elite groups existing in the United States of America. The term commonly includes the so-called "blue bloods" (multi-generational wealth combined with leadership of high society) such as the Astor or Roosevelt families. There is disagreement over whether the "nouveau riche" should be included as members of the upper class or whether this term should exclusively be used for established families. Twentieth century sociologist W. Lloyd Warner divided the upper class into two sections: the "upper-upper class" (or bourgeoise) and "lower-upper class" (or yuppies). The former includes established upper-class families while the latter includes those with great wealth. As there is no defined lower threshold for the upper class it is difficult, if not outright impossible, to determine the exact number or percentage of American households that could be identified as being members of the upper-class(es).

Income and wealth statistics may serve as a helpful guideline as they can be measured in a more objective manner. In 2005, approximately one and half percent (1.5%) of households in the United States had incomes exceeding $250,000 with the top 5% having incomes exceeding $157,000. Furthermore only 2.6% of household held assets (excluding home equity) of more than one-million dollars. One could therefore fall under the assumption that less than five percent of American society are members of rich households. The richest 1% of the American population owns as much as the combined wealth of the bottom 90%, or perhaps even more.

Members of the upper class control and own significant portions of the corporate America and may exercise indirect power through the investment of capital. The high salaries and the potential for amassing great wealth through stock options have greatly increased for the power and visibility of the "corporate elite". Many sociologists and commentators, however, make a distinction between the upper class (in the sense of those in the families of inherited wealth) and the corporate elite. By implication, the upper class is held in lower regard (as inheritors of idle wealth) than the self-made millionaires in prestigious occupations.

Yet another important feature of the upper class is that of inherited privilege. While most Americans, including those in the upper-middle class need to actively maintain their status, upper class persons do not need to work in order to maintain their status. Status tends to be passed on from generation to generation without each generation having to re-certify its status. Overall, the upper class is the financially best compensated and one of the most influential socio-economic classes in American society.

Corporate Elite

The high salaries and, especially, the potential wealth through stock options, has supported the term corporate elite. Top executives, including Chief Executive Officers, are among the financially best compensated occupations in the United States. The median annual earnings for a CEO in the United States were $140,350 (exceeding the income of more than 90% of U.S. households). The Wall Street Journal reports the median compensation for CEOs of 350 major corporations was $6,000,000 in 2005 with most of the money coming from stock options. In New York City in 2005, the median income (including bonuses) of a corporate "chief operating officer" (the #2 job) was $377,000. The total compensation for a "top IT officer" in charge of information technology in New York City was $218,000. Thus even below the CEO level of top corporations, financial compensation will usually be sufficient to propel a households with a mere one income earner in the top 1%. In 2005 only 1.5% of American households had incomes above $250,000 with many reaching this level only through having two income earners.

> *"Top executives are among the highest paid workers in the U.S. economy. However, salary levels vary substantially depending on the level of managerial responsibility; length of service; and type, size, and location of the firm. For example, a top manager in a very large corporation can earn significantly more than a counterpart in a small firm.*
>
> *Median annual earnings of general and operations managers in May 2004 were $77,420. The middle 50% earned between $52,420 and $118,310. Because the specific responsibilities of general and operations managers vary significantly within industries, earnings also tend to vary considerably...the Median annual earnings of chief executives in May 2004 were $140,350; although chief executives in some industries earned considerably more...the median income of chief executive officers in the nonprofit sector was $88,006 in 2005, but some of the highest paid made more than $700,000."*

—U.S. Department of Labour.

Many politically powerful people make money before coming to office, but in general the political power elite have official incomes in the $150,000 to $185,000 range; members of Congress are paid $165,000, and are effectively required to have a residence in their district as well as one in Washington.

Upper Middle

The upper middle class consists of highly-educated salaried professionals whose work is largely self-directed. Many have advanced graduate degrees and household incomes commonly exceed the high five-figure range. Members of this class commonly value higher education-most holding advanced academic degrees-and are often involved with personal and professional networks including professional organizations. The upper middle class tends to have great influence over the course of society.

Occupations which require high educational attainment, are well-compensated and are held in high public esteem such as physicians, lawyers, scientists and professors are largely considered to be upper middle class. The very well-educated are seen as trend setters with movements such as the anti-smoking movement, pro-fitness movement, organic food movement, and environmentalism being largely indigenous to this particular socio-economic grouping. Education serves as perhaps the most important value and also the most dominant entry barrier of the upper middle class.

Sociologists such as Dennis Gilbert, Willam Thompson, and Joseph Hickey estimate the upper middle class to constitute roughly 15% of the population (or roughly three in every twenty persons). The main hallmark and most distinguishing feature of this class is its high educational attainment.

Middle Class

The middle class is perhaps the mostly vaguely defined of the social classes. The term can be used either to describe a relative elite of professionals and managers – also called the upper middle class – or it can be used to describe those in-between the extremes of wealth, disregarding considerable differences in income, culture, educational attainment, influence, and occupation. As with all social classes in the U.S., there are no definite answers as to what is and what is not middle class. Sociologists such as Dennis Gilbert, James Henslin, William Thompson, and Joseph Hickey have brought forth class models in which the middle class is divided into two sections that combined constitute 47% to 49% of the population. The upper middle or professional class constitutes the upper end of the middle class which consists of highly educated, well-paid professionals with considerable work autonomy. The lower end of the middle class – called either lower middle class or just middle class – consists of semi-professionals, craftsmen,office staff, and sales employees who often have college degrees and are very loosely supervised.

Jerry Carrier in "The Making of the Slave Class" asserts that American's perception of a middle class country is a myth. He asserts that about two-thirds of American's are working class and their incomes and class status are severely limited because of culture, particularly Christian culture. He contends that no real class mobility exists in America.

> *"Everyone wants to believe they are middle class. For people on the bottom and the top of the wage scale the phrase connotes a certain Regular Joe cachet. But this eagerness to be part of the group has led the definition to be stretched like a bungee cord."*
>
> —Dante Chinni, the Christian Science Monitor,

Although income thresholds cannot be determined since social classes lack distinct boundaries and tend to overlap, sociologists and economist have put forward certain income figures they find indicative of middle class households. Sociologist Leonard Beeghley identifies a husband making roughly \$57,000 and a wife making roughly \$40,000 with a household income of roughly \$97,000 as a typical middle class family. Sociologists William Thompson and Joseph Hickey identify household incomes between \$35,000 and \$75,000 as typical for the lower middle and \$100,000 or more as typical for the upper middle class. Though it needs to be noted that household income distribution neither reflects standard of living nor class status with complete accuracy.

Traditional Middle Class

Those households more or less at the centre of society may be referred to as being part of the American middle or middle-middle class in vernacular language use. In the academic models featured in this article, however, the middle class does not constitute a strong majority of the population. Those in the middle of the socio-economic strata—the proverbial Average Joe—are commonly in the area where the working and lower middle class overlap.

The most prominent academic models split the middle class into two sections. Yet, it remains common for the term middle class to be applied for anyone in between either extreme of the socio-economic strata. The middle class is then often sub-divided into an upper-middle, middle-middle, and lower-middle class. In colloquial descriptions of the class system the middle-middle class may described as consisting of those in the middle of the social strata. Politicians and television personalities such as Lou Dobbs can be seen using the term middle class in this manner, especially when discussing the middle class squeeze. The wide discrepancy between the academic models and public opinions that lump highly educated professionals together in the same class with secretaries, may lead to the conclusion that public opinion on the subject has become largely ambiguous.

Lower Middle Class

The lower middle class is, as the name implies, generally defined as those less privileged than the middle class. People in this class commonly work in supporting occupations. Although they seldom hold advanced academic degrees, a college degree (usually a bachelor's degree) is almost always required for entry into the lower middle class.

Sociologists Dennis Gilbert, William Thompson, and Joseph Hickey, however, only divide the middle class into two groups. In their class modes the middle class only consists of an upper and lower middle class. The upper middle class, as described above, constitutes roughly 15% of the population with highly educated white collar professionals who commonly have salaries in the high 5-figure range and household incomes in the low six figure range. Semi-professionals with Bachelor's degrees and some college degrees constitute the lower middle class. Their class models show the lower middle class positioned slightly above the middle of the socio-economic strata. Those in blue and pink collar as well as clerical occupations are referred to as working class in these class models.

Working Class

The term working class is applies to those that work at this tier in the social hierarchy. Definitions of this term vary greatly. While Lloyd Warner found the vast majority of the American population to be in either the upper-lower class or lower-lower class in 1949, modern-day experts such as Michael Zweig, an economist for SUNY–Stony Brook, argue that the working class constitutes most of the population.

Dennis Gilbert places 13% of households among the "working poor" with 12% being in the "underclass". Thompson & Hickey place roughly 17% to 20% of households in the lower classes. The lower classes constituting roughly a fifth to a quarter of American society consists mainly of low-rung retail and service workers as well as the frequently unemployed and those not able to work. Overall, 13% of the population fall below the poverty threshold. Hunger and food insecurity were present in the lives of 3.9% of American households, while roughly twenty-five million Americans (ca. 9%) participated in the food stamp program.

Agriculture

Agriculture is a major industry in the United States and the country is a net exporter of food. As of the last census of agriculture in 2007, there were 2.2 million farms, covering an area of 922 million acres (373 million hectares), an average of 418 acres (170 hectares) per farm History.

Corn, turkeys, tomatoes, potatoes, peanuts, and sunflower seeds constitute some of the major hold overs from the agricultural endowment of the Americas.

European agriculture practices greatly affected the New England landscape, leaving behind many physical foot prints. Colonists brought livestock over from Europe which caused many changes to the land. Grazing animals required a lot of land and food to sustain them and due to grazing, native grasses were destroyed and European species began to replace them. New species of weeds were introduced and began to thrive as they were capable of withstanding the grazing of animals, whereas native species could not.

Along with livestock changing the plant species in New England from the original native species to European species they also contributed to the deterioration of the forests and fields. Colonists would cut down the trees and then allow their cattle and livestock to graze freely in the forest and never plant more trees. The animals trampled and tore up the ground so much as to cause long-term destruction and damage.

Soil exhaustion was a huge problem in New England agriculture. Plowing with oxen did allow the colonist to farm more land but it increased erosion and decreased soil fertility. This was due to deeper plow cuts in the soil that allowed the soil more contact with oxygen causing nutrient depletion. In grazing fields, the large number of cattle in the New England, the soil was being compacted by the cattle and this didn't give the soil enough oxygen to sustain life.

In the U.S., farms spread from the colonies westward along with the settlers. In cooler regions, wheat was often the crop of choice when lands were newly settled, leading to a "wheat frontier" that moved westward over the course of years. Also very common in the antebellum Midwest was farming corn while raising hogs, complementing each other especially since it was difficult to get grain to market before the canals and railroads. After the "wheat frontier" had passed through an area, more diversified farms including dairy cattle generally took its place. Warmer regions saw plantings of cotton and herds of beef cattle. In the early colonial south, raising tobacco and cotton was common, especially through the use of slave labour until the Civil War. In the northeast, slaves were used in agriculture until the early 19th century. In the Midwest, slavery was prohibited by the Freedom Ordinance of 1787.

The introduction and broad adoption of scientific agriculture since the mid nineteenth century has made a large improvement in the USA's economic growth. This development was facilitated by the Morrill Act and the Hatch Act of 1887 which established in each state a land-grant university (with a mission to teach and study agriculture) and a federally-funded system of agricultural experiment stations and cooperative extension networks which place extension agents in each state.

Soybeans were not widely cultivated in the United States until the 1950s, when soybeans began to replace oats and wheat.

Significant areas of farmland were abandoned during the Great Depression and incorporated into nascent national forests. Later, "Sodbuster" and "Swampbuster" restrictions written into federal farm programs starting in the 1970s reversed a decades-long trend of habitat destruction that began in 1942 when farmers were encouraged to plant all possible land in support of the war effort. In the United States, federal programs administered through local Soil and Water Conservation Districts provide technical assistance and partial funding to farmers who wish to implement management practices to conserve soil and limit erosion.

Major Agricultural Products

1.	Corn	256,900,000
2.	Cattle meat	11,736,000
3.	Cow's milk, whole, fresh	78,155,000
4.	Chicken meat	15,006,000
5.	Soybeans	65,800,000
6.	Pig meat	8,574,000
7.	Wheat	63,590,000
8.	Cotton lint	3,968,000
9.	Hen eggs	5,141,000
10.	Turkey meat	2,584,000
11.	Tomatoes	12,275,000
12.	Potatoes	20,820,000
13.	Grapes	6,126,000
14.	Oranges	10,473,000

15.	Rice, paddy	9,034,000
16.	Apples	4,242,000
17.	Sorghum	10,446,000
18.	Lettuce	4,490,000
19.	Cottonseed	6,073,000
20.	Sugar beets	27,760,000

The only other crops to ever appear in the top 20 in the last 40 years were, commonly, tobacco, barley, and oats, and, rarely, peanuts, almonds, and sunflower seeds (in all, only 26 of the 188 crops the FAO tracks worldwide). Alfalfa and hay would both be in the top ten in 2003 if they were tracked by FAO.

Crops

Value of Production

Major Crops in the U.S.A.-1997 (in US$ Billions)

Corn	24.4
Soybeans	17.7
Wheat	8.6
Alfalfa	8.3
Cotton	6.1
Hay, other than alfalfa	5.1
Tobacco	3.0
Rice	1.7
Sorghum	1.4
Barley	.9

Note alfalfa and hay are not tracked by the FAO and the production of tobacco in the U.S. has fallen 60% between 1997 and 2003.

Yield

U.S. agriculture has a high yield relative to other countries. The yield was (in 2004):

- Corn for grain, average of 160.4 bushels harvested per acre (10.07 t/ha)

- Soybean for beans, average of 42.5 bushels harvested per acre (2.86 t/ha)
- Wheat, average of 43.2 bushels harvested per acre (2.91 t/ha, was 44.2 bu/ac or 2.97 t/ha in 2003).

Livestock

The major livestock industries in the United States are:

- Dairy Cattle
- Beef Cattle
- Swine (also called hogs or pigs)
- Poultry
- Sheep.

Inventories in the United States at the end of 1997 were absolutly:

• Chickens	403,000,000
• Cattle	99,500,000
• Hogs	59,900,000
• Sheep	7,600,000

Goats, horses, turkeys and bees are also raised, though in lesser quantities. Inventory data is not as readily available as for the major industries.

For the three major goat-producing states (AZ, NM, and TX) there were 1,200,000 goats at the end of 2002. There were 5,300,000 horses in the United States at the end of 1998. There were 2,500,000 colonies of bees at the end of 2002.

Farm Type or Majority Enterprise Type

Farm type is based on which commodities are the majority crops grown on a farm. Nine common types include:

- Cash Grains includes corn, soybeans and other grains (wheat, oats, barley, sorghum), dry edible beans and peas, and rice.
- Tobacco
- Cotton

- Other Field Crops includes peanuts, potatoes, sunflowers, sweet potatoes, sugarcane, broomcorn, popcorn, sugar beets, mint, hops, seed crops, hay, silage, forage, etc. Tobacco and cotton can be included here if not
- High Value Crops includes fruits, vegetables, melons, tree nuts, greenhouse and nursery crops, and horticultural specialties.
- Cattle
- Hogs
- Dairy
- Poultry and Eggs.

Governance

Agriculture in the United States is primarily governed by periodically-renewed U.S. farm bills. Governance is both a federal and a local responsibility with the United States Department of Agriculture being the federal department responsible.

Pro-agriculture Americans form an extremely powerful interest group in American politics and have since the founding of the USA. Government aid includes research into crop types and regional suitability as well as many kinds of subsidies, some price supports and loan programs. U.S. farmers are not subject to production quotas and some laws are different for farms compared to other workplaces.

Labour laws prohibiting children in other workplaces provide some exemptions for children working on farms with complete exemptions for children working on their family's farm. Children can also gain permits from vocational training schools or the 4-H club which allow them to do jobs they would otherwise not be permitted to do.

A large part of the U.S. farm workforce is made up of migrant and seasonal workers, many of them recent immigrants from Latin America or aliens working under work permits. Additional laws apply to these workers and their housing which is often provided by the farmer.

Employment

In 1870, 70-80 percent of the US population was employed in agriculture. As of 2008, approximately 2-3 percent of the population is directly employed in agriculture. See this USDA link, http://www.csrees.usda.gov/qlinks/extension.html for more information and history on U.S. agriculture.

In 2004, of the 145 million employed workers in the US, 834,000 of them held jobs as agricultural workers. 83% of these jobs were as farm workers. The median hourly income as of May 2004 was $7.70 for farmworkers planting, growing and harvesting crops, and $8.31 for farmworkers tending to animals.

Agriculture Safety and Health

Agriculture ranks among the most hazardous industries. Farmers are at high risk for fatal and nonfatal injuries, work-related lung diseases, noise-induced hearing loss, skin diseases, and certain cancers associated with chemical use and prolonged sun exposure. Farming is one of the few industries in which the families (who often share the work and live on the premises) are also at risk for injuries, illness, and death. In an average year, 516 workers die doing farm work in the U.S. (1992-2005). Of these deaths, 101 are caused by tractor overturns. Every day, about 243 agricultural workers suffer lost-work-time injuries, and about 5% of these result in permanent impairment.

Agriculture is the most dangerous industry for young workers, accounting for 42% of all work-related fatalities of young workers in the U.S. between 1992 and 2000. Unlike other industries, half the young victims in agriculture were under age 15. For young agricultural workers aged 15–17, the risk of fatal injury is four times the risk for young workers in other workplaces Agricultural work exposes young workers to safety hazards such as machinery, confined spaces, work at elevations, and work around livestock.

An estimated 1.26 million children and adolescents under 20 years of age resided on farms in 2004, with about 699,000 of these youth performing work on the farms. In addition to the youth who live on farms, an additional 337,000 children and adolescents were hired to work on U.S. farms in 2004. On average, 103 children are killed annually on farms (1990-1996). Approximately 40 percent of these deaths were work-related. In 2004, an estimated 27,600 children and adolescents were injured on farms; 8,100 of these injuries were due to farm work.

Farmworkers

Before industrialization, "yeoman farmers"—self-sufficient, politically independent landowners—made up a large portion of the country's population. Jeffersonian democracy and Jacksonian democracy successfully expanded the political rights of the yeomen, and the geographical extent of the nation to provide them farms. This culminated in the Homestead Act of 1862 which provided hundreds of thousands of free farms. Before 1865 large southern plantations used slaves. After emancipation, a system of sharecropping and tenant farming for both whites and blacks in the South provided a semi-independent status for farmers who did not own their land. In contemporary times, migrant agricultural workers—mostly Mexicans—perform field and packing work.

Farmers

Only 0.7% of the population of the United States is employed in the agricultural sector. Most are proprietors of independent farms. Once the dominant American social class, this group diminished in overall numbers during the 20th century, as farm holdings grew more consolidated, farming operations became more mechanized, and most of the population migrated to urban areas.

Today, the agricultural sector has essentially taken on the characteristics of business and industry generally. In contemporary usage, a "farmer" is someone who owns and operates a farm, which more often than not will be a sizable business enterprise; "agricultural workers" or "farm workers", who perform the actual work associated with farming, typically come out of the lower classes; indeed, they are often near-destitute immigrants or migrant farm workers. In this respect, farming mirrors big business: like any enterprise, a farm has owners (who may be a family or a corporation), salaried managers, supervisors, foremen and workers.

With the number of farms steadily diminishing, the stereotypical humble homestead is increasingly the exception, for viable farming now means agribusiness; the large amounts of capital required to operate a competitive farm require large-scale organization. The large landowners in California's Central Valley, Coachella Valley and Imperial Valley fall squarely within the upper class. Among farmers, "income" in the conventional sense is not an accurate standard of wealth measurement, because farmers typically keep their official income low by placing their assets into farming corporations rather than drawing the money directly. The stereotypical poor, marginal farmer "eking out a living" from the soil, an image deeply ingrained in most Americans' minds by folklore, films, and even history texts, has now been largely displaced by agribusiness, which has bought them out and consolidated their holdings.

Class Mobility

Class ascendancy—namely that each successive generation will have a higher standard of living than its predecessor—is a central theme in American literature and culture and plays a key role in the American dream. While social class in the United States is thought to be largely based on achievement, climbing the social ladder is more difficult for those born into less advantageous positions.

Occupation (perhaps the most important class component), educational attainment, and income can be increased through a lifetime. However, factors such as wealth inheritance and local education system—which often provides lower quality education to those in poor school districts—may make rising out of poverty a challenge. Class mobility in the United States decreased between the 1970s and the 1990s.

Recently a number of works have denied the existence of upward mobility. The New York Times in 2005, in a series of articles called "Class Matters" shed doubt on upward mobility and in 2010 Jerry Carrier in his book "The Making of the Slave Class stated flatly there is no upward mobility in America and that the reason for this is culture which limits opportunity and income.

Class and Politics

Income remains one of the main indicators of class, as it commonly reflects high educational attainment as well as a prestigious occupation. Today there remains a strong correlation, if not causation, between political activism as well as political representation and household income. Furthermore, a discrepancy in the political attitudes can be found among individuals residing in households with differing incomes.

During the 2000 election, voter turnout among those in the top 26% with household incomes exceeding $75,000 was 27% higher than the average. Some rather controversial issues such as race were impacted by household income. With higher voter turnout and affiliation with professional organizations and other popular political lobbies such as the AARP or NAACP, those with higher incomes were more likely to be represented in the political process with the government being more attuned to their needs.

Education also has an effect on voting record. Generally, the more educated a person is, the more likely he or she is to vote with those having graduate degrees being more likely to vote Democrat. While a rise in income generally increases the likelihood of a person voting Republican, many upper middle class professionals with post-graduate degrees are more likely to vote Democrat.

The presence of upper middle class liberals as households with incomes exceeding $100,000 and those with graduate degrees were the only demographics where Ralph Nader won 1% of the vote. Even though those with post-graduate degrees only constituted 8.9% of the general population, they constituted 16% of all voters. This means that the most highly educated 16% of voters were expected to vote Democrat even though most of their similarly high paid but lesser educated counterparts did not.

Class and Health

Income also had a significant impact on health as those with higher incomes had better access to health care facilities, higher life expectancy, lower infant mortality rate and increased health consciousness. While the United States lacks socialized medicine similar to that found in many other post-industrialized developed nations across Europe and Asia, 85% of the U.S. population were insured in 2005. Yet, discrepancies seem to remain beyond the difference between insured and uninsured. In 2006, Harvard researchers divided the U.S. into "eight Americas."

Life expectancy ranges from 84.9 years for the 10,400,000 Asian Americans who had an average per capita income of $21,566. Urban African Americans with an average per capita income of a $14,800 had a life expectancy of merely 71.1 years, although this does not control for natural longevity differences among race.

Furthermore, the United States like other post-industrial nations saw increased health consciousness among persons of higher social status. Persons of higher status are less likely to smoke, more likely to exercise regularly and be more conscious of their diet. Additionally, poor Americans are more likely to consume lower quality, processed foods. One can therefore conclude that low socio-economic status contributes to a person's likelihood of being obese.

Popular Media

The world's first commercial motion picture exhibition was given in New York City in 1894, using Thomas Edison's Kinetoscope. The next year saw the first commercial screening of a projected film, also in New York, and the United States was in the forefront of sound film's development in the following decades. Since the early 20th century, the U.S. film industry has largely been based in and around Hollywood, California. Director D. W. Griffith was central to the development of film grammar and Orson Welles's Citizen Kane (1941) is frequently cited as the greatest film of all time. American screen actors like John Wayne and Marilyn Monroe have become iconic figures, while producer/entrepreneur Walt Disney was a leader in both animated film and movie merchandising. The major film studios of Hollywood have produced the most commercially successful movies in history, such as Star Wars (1977) and Titanic (1997), and the products of Hollywood today dominate the global film industry. Americans are the heaviest television viewers in the world, and the average viewing time continues to rise, reaching five hours a day in 2006. The four major broadcast networks are all commercial entities. Americans listen to radio programming, also largely commercialized, on average just over two-and-a-half hours a day. Aside from web portals and search engines, the most popular websites are Facebook, YouTube, Wikipedia, Blogger, eBay, and Craigslist.

The rhythmic and lyrical styles of African American music have deeply influenced American music at large, distinguishing it from European traditions. Elements from folk idioms such as the blues and what is now known as old-time music were adopted and transformed into popular genres with global audiences.

Jazz was developed by innovators such as Louis Armstrong and Duke Ellington early in the 20th century. Country music developed in the 1920s, and rhythm and blues in the 1940s. Elvis Presley and Chuck Berry were among the mid-1950s pioneers of rock and roll. In the 1960s, Bob Dylan emerged from the folk revival to become one of America's most celebrated songwriters and James Brown led the development of funk. More recent American creations include hip hop and house music. American pop stars such as Presley, Michael Jackson, and Madonna have become global celebrities.

Literature, Philosophy and the Arts

In the eighteenth and early nineteenth centuries, American art and literature took most of its cues from Europe. Writers such as Nathaniel Hawthorne, Edgar Allan Poe, and Henry David Thoreau established a distinctive American literary voice by the middle of the 19th century. Mark Twain and poet Walt Whitman were major figures in the century's second half; Emily Dickinson, virtually unknown during her lifetime, is now recognized as an essential American poet. A work seen as capturing fundamental aspects of the national experience and character—such as Herman Melville's Moby-Dick (1851), Twain's The Adventures of Huckleberry Finn (1885), and F. Scott Fitzgerald's The Great Gatsby (1925)—may be dubbed the "Great American Novel."

Eleven U.S. citizens have won the Nobel Prize in Literature, most recently Toni Morrison in 1993. Ernest Hemingway, the 1954 Nobel laureate, is often named as one of the most influential writers of the 20th century. Popular literary genres such as the Western and hardboiled crime fiction developed in the United States. The Beat Generation writers opened up new literary approaches, as have postmodernist authors such as John Barth, Thomas Pynchon, and Don DeLillo.

The transcendentalists, led by Thoreau and Ralph Waldo Emerson, established the first major American philosophical movement. After the Civil War, Charles Sanders Peirce and then William James and John Dewey were leaders in the development of pragmatism. In the 20th century, the work of W. V. O. Quine and Richard Rorty, built upon by Noam Chomsky, brought analytic philosophy to the fore of U.S. academics. John Rawls and Robert Nozick led a revival of political philosophy.

In the visual arts, the Hudson River School was a mid-19th-century movement in the tradition of European naturalism. The 1913 Armory Show in New York City, an exhibition of European modernist art, shocked the public and transformed the U.S. art scene. Georgia O'Keeffe, Marsden Hartley, and others experimented with new styles, displaying a highly individualistic sensibility. Major artistic movements such as the abstract expressionism of Jackson Pollock and Willem de Kooning and the pop art of Andy Warhol and Roy Lichtenstein developed largely in the United States. The tide of modernism and then postmodernism has brought fame to American architects such as Frank Lloyd Wright, Philip Johnson, and Frank Gehry.

One of the first major promoters of American theater was impresario P. T. Barnum, who began operating a lower Manhattan entertainment complex in 1841. The team of Harrigan and Hart produced a series of popular musical comedies in New York starting in the late 1870s. In the 20th century, the modern musical form emerged on Broadway; the songs of musical theater composers such as Irving Berlin, Cole Porter, and Stephen Sondheim have become pop standards. Playwright Eugene O'Neill won the Nobel literature prize in 1936; other acclaimed U.S. dramatists include multiple Pulitzer

Prize winners Tennessee Williams, Edward Albee, and August Wilson.

Though largely overlooked at the time, Charles Ives's work of the 1910s established him as the first major U.S. composer in the classical tradition; other experimentalists such as Henry Cowell and John Cage created an American approach to classical composition. Aaron Copland and George Gershwin developed a unique synthesis of popular and classical music. Choreographers Isadora Duncan and Martha Graham helped create modern dance, while George Balanchine and Jerome Robbins were leaders in 20th century ballet. Americans have long been important in the modern artistic medium of photography, with major photographers including Alfred Stieglitz, Edward Steichen, and Ansel Adams. The newspaper comic strip and the comic book are both U.S. Innovations. Superman, the quintessential comic book superhero, has become an American icon.

Food

Mainstream American culinary arts are similar to those in other Western countries. Wheat is the primary cereal grain. Traditional American cuisine uses ingredients such as turkey, white-tailed deer venison, potatoes, sweet potatoes, corn, squash, and maple syrup, indigenous foods employed by Native Americans and early European settlers. Slow-cooked pork and beef barbecue, crab cakes, potato chips, and chocolate chip cookies are distinctively American styles. Soul food, developed by African slaves, is popular around the South and among many African Americans elsewhere. Syncretic cuisines such as Louisiana creole, Cajun, and Tex-Mex are regionally important.

Characteristic dishes such as apple pie, fried chicken, pizza, hamburgers, and hot dogs derive from the recipes of various immigrants. French fries, Mexican dishes such as burritos and tacos, and pasta dishes freely adapted from Italian sources are widely consumed. Americans generally prefer coffee to tea. Marketing by U.S. industries is largely responsible for making orange juice and milk ubiquitous breakfast beverages. During the 1980s and 1990s, Americans' caloric intake rose 24%; frequent dining at fast food outlets is associated with what health officials call the American "obesity epidemic." Highly sweetened soft drinks are widely popular; sugared beverages account for 9% of the average American's caloric intake.

Sports

Since the late 19th century, baseball has been regarded as the national sport; American football, basketball, and ice hockey are the country's three other leading professional team sports. College football and basketball attract large audiences. Football is now by several measures the most popular spectator sport. Boxing and horse racing were once the most watched individual sports, but they have been eclipsed by golf and auto racing, particularly NASCAR. Soccer is played widely at the youth and amateur levels. Tennis and many outdoor sports are popular as well.

While most major U.S. sports have evolved out of European practices, basketball, volleyball, skateboarding, snow boarding, and cheerleading are American inventions. Lacrosse and surfing arose from Native American and Native Hawaiian activities that predate Western contact. Eight Olympic Games have taken place in the United States. The United States has won 2,301 medals at the Summer Olympic Games, more than any other country, and 253 in the Winter Olympic Games, the second most.

Measurement Systems

The country retains United States customary units, constituted largely by British imperial units such as yards, miles, and degrees Fahrenheit. Distinct units include the U.S. gallon and pint volume measurements. The United States is one of three countries, along with Burma and Liberia, that has not officially adopted the metric system. However, metric units are increasingly used in science, medicine, and many industrial fields.

9

New England—A Region of Limited Resources

New England is a region in the northeastern corner of the United States, bordered by the Atlantic Ocean, Canada and the state of New York, consisting of the states of Maine, New Hampshire, Vermont, Massachusetts, Rhode Island, and Connecticut.

In one of the earliest European settlements in the New World, Pilgrims from the Kingdom of England first settled in New England in 1620, in the colony of Plymouth. Ten years later, the Puritans settled north of Plymouth Colony in Boston, thus forming Massachusetts Bay Colony in 1630. In the late 18th century, the New England colonies would be among the first North American British colonies to demonstrate ambitions of independence from the British Crown via the American Revolution, although they would later oppose the War of 1812 between the United States and United Kingdom of Great Britain and Ireland.

New England was known to have produced the first pieces of American literature and philosophy and was home to the beginnings of free public education. In the 19th century, it played a prominent role in the movement to abolish slavery in the United States. It was the first region of the United States to be transformed by the Industrial Revolution.

The region is one of the most liberal in the United States and it voted for the Democratic Party Presidential nominee in the 1992, 1996, 2004, and 2008 elections, and every state but New Hampshire voted for Al Gore in the presidential election of 2000. Currently all members of the United States House of Representatives from New England belong to or caucus with the Democratic Party. The only democratic socialist in the United States Congress is from New England. New England has also produced the only current independent senators-Bernie Sanders representing Vermont and Joseph Lieberman representing Connecticut.

History

New England's earliest inhabitants were Algonquian-speaking Native Americans including the Abenaki, the Penobscot, and the Wampanoag. Prior to the arrival of Europeans, the Western Abenakis inhabited New Hampshire and Vermont, as well as parts of Quebec and western Maine. Their principal town was Norridgewock, in present-day Maine. The Penobscot were settled along the Penobscot River in Maine. The Wampanoag occupied southeastern Massachusetts, Rhode Island, and the islands of Martha's Vineyard and Nantucket. The Connecticut region was inhabited by the Mohegan and Pequot tribes prior to European colonization.

The Virginia Companies Compete

On April 10, 1606, King James I of England issued two charters, one each for the Virginia Companies, of London and Plymouth, respectively. Due to a duplication of territory (between Chesapeake Bay and Long Island Sound), the two companies were required to maintain a separation of 100 Miles, even where the two charters overlapped.

These were privately-funded proprietary ventures, and the purpose of each was to claim land for England, trade, and return a profit. Competition between the two companies grew to where their potential New World territory overlapped, and would be finalized based upon results.

The London Company was authorized to make settlements from North Carolina to New York (31 to 41 degrees North Latitude), provided there was no conflict with the Plymouth Company's charter.

The Popham Colony was planted at the mouth of Maine's Kennebec River by the Virginia Company of Plymouth in the fall of 1607. Unlike the Jamestown Settlement, it was not successful, and was abandoned the following spring. The Virginia Company of Plymouth's charter included land extending as far as present-day northern Maine. Captain John Smith, exploring the shores of the region in 1614, named the region "New England" in his account of two voyages there, published as A Description of New England.

Plymouth Council for New England

The name "New England" was officially sanctioned on November 3, 1620, when the charter of the Virginia Company of Plymouth was replaced by a royal charter for the Plymouth Council for New England, a joint stock company established to colonize and govern the region.

Shortly afterwards, in December 1620, a permanent settlement was established near present-day Plymouth by the Pilgrims, English religious separatists arriving via Holland, after they famously disembarked at Plymouth Rock. The Massachusetts Bay Colony, which would come to dominate the area, was established in 1628 with its major city of Boston established in 1630.

Banished from Massachusetts for heresy, Roger Williams led a group south, and founded Providence, Rhode Island in 1636. On March 3 of the same year, Thomas Hooker also left Massachusetts and the Connecticut Colony was granted a charter, establishing its own government in Hartford. At this time, Vermont was yet unsettled, and the territories of New Hampshire and Maine were governed by Massachusetts.

New England Confederation

In these early years, relationships between colonists and Native Americans alternated between peace and armed skirmishes. Six years after the bloodiest of these, the Pequot War in 1643, the colonies of Massachusetts Bay, Plymouth, New Haven, and Connecticut joined together in a loose compact called the New England Confederation (officially "The United

Colonies of New England"). The confederation was designed largely to coordinate mutual defence against possible wars with Native Americans, the Dutch in the New Netherland colony to the west, the Spanish in the south, and the French in New France to the north, as well as to assist in the return of runaway slaves. The confederation lost its influence when Massachusetts refused to commit itself to a war against the Dutch.

The first coins struck in the Colonies, prompted by a shortage of change, were the New England coins produced by the Massachusetts Bay Colony. The first series was a simple design including "NE" on the obverse and the various denominations on the reverse. Other series included the "Willow," "Oak," and "Pine Tree." The "Pine Tree" coinage was the last type in the series, struck by coiner John Hull. Although the majority were dated 1652, it is generally acknowledged that production spanned about thirty years, despite the disapproval of King Charles II.

All Puritans worked. As a result there was a "Working ethic." Workers were the backbone of the community. There were blacksmiths, wheelrights, carpenters, joiners, cordwainers, tanners, ironworkers, spinners, and weavers, when someone needed something-unlike the Southern colonies, who had to buy these items from England.

Dominion of New England

In 1686, King James II, concerned about the increasingly independent ways of the colonies, including their self-governing charters, open flouting of the Navigation Acts, and increasing military power, established the Dominion of New England, an administrative union comprising all of the New England colonies. On August 11, 1688, the provinces of New York and New Jersey, seized from the Dutch in 1664, and confirmed on September 12, 1673, were added. The union, imposed from the outside and contrary to the rooted democratic tradition of the region, was highly unpopular among the colonists.

Nevertheless, those two present states are reckoned as "greater New England" in a social or cultural context, as that is where Yankee colonists expanded to; before 1776. Cultural identity in that era changed once one moved to Pennsylvania, as the Pennamite-Yankee War attests to. Colonists from New England proper in that era, were rather well received in the Mohawk Valley and on Long Island in New York.

After the Glorious Revolution in 1689, Bostonians imprisoned the Royal Governor and other sympathizers of King James II on April 18, 1689, thus ending the Dominion Of New England de facto. The charters of the colonies were significantly modified after this change in English politics, with the appointment of Royal Governors to nearly every colony.

An uneasy tension existed between the Royal Governors, their officers, and the elected governing bodies of the colonies. The governors wanted unlimited authority, and the different layers of locally elected officials would often resist them. In most cases, the local town governments continued operating as self-governing bodies, just as they had before the appointment of the Royal Governors. This tension culminated itself in the American Revolution, boiling over with the break-out of the American War of Independence in 1775.

Region of the United States

The colonies were now formally united as newly-formed states in a larger (but not yet federalist) union United States of America.

In the 18th century and the early 19th century, New England was still considered to be a very distinct region of the colony and country, as it is today.

During the War of 1812, there was a limited amount of talk of secession from the Union, as New England merchants, just getting back on their feet, opposed the war with their greatest trading partner—Great Britain. Delegates from all over New England met in Hartford in the winter of 1814-15. The gathering was called the Hartford Convention. The 27 delegates met to discuss changes to the US Constitution that would protect the region from similar legislation and attempt to keep political power in the region.

For the remainder of the Antebellum period, New England remained distinct. In terms of politics, it often went against the grain of the rest of the country. Massachusetts and Connecticut were among the last refuges of the Federalist Party, and, when the Second Party System began in the 1830s, New England became the strongest bastion of the new Whig Party.

The Whigs were usually dominant throughout New England, except in the more Democratic Maine and New Hampshire. Many of the leading statesmen — including Daniel Webster — hailed from the region. New England was also distinct in other ways. It was, as a whole, the most urbanized part of the country (the 1860 Census showed that 32 of the 100 largest cities in the country were in New England), as well as the most educated. Many of the major literary and intellectual figures produced by the United States in the Antebellum period were New Englanders, including Ralph Waldo Emerson, Henry David Thoreau, Nathaniel Hawthorne, Henry Wadsworth Longfellow, John Greenleaf Whittier, George Bancroft, William H. Prescott, and others.

New England was also an early centre of the industrial revolution. Pawtucket, Rhode Island is considered the birthplace of America's industrial revolution, the city in which Slater's Mill was founded.

Despite the nickname, several textile mills were already underway before Slater Mill was established. The first textile mill in the United States was built in 1787 at Beverly, Massachusetts by entrepreneur John Cabot. Towns like Lawrence, Massachusetts, Lowell, Massachusetts, Woonsocket, Rhode Island, and Lewiston, Maine became famed as centres of the textile industry following models from Slater Mill and Beverly.

The textile manufacturing in New England was growing rapidly, which caused a shortage of workers. Recruiters were hired by mill agents to bring young women from the countryside to work in the factories. Between 1830 and 1860, thousands of farm girls came from their rural homes in New England to work in the mills. Farmers' daughters left their homes to aid their families financially, save for marriage, and widen their horizons. They also left their homes due to population pressures to look for opportunities in expanding New England cities. Stagecoach and railroad services made it easier for the rapid flow of workers to travel from the country to the city. The majority of female workers came from rural farming towns in northern New England.

As the textile industry grew, immigration grew as well. As the number of Irish workers in the mills increased, the number of young women working in the mills decreased. Mill employment of women caused a population boom in urban centres.

New England and areas settled from New England, like Upstate New York, Ohio's Western Reserve and the upper midwestern states of Michigan and Wisconsin, also proved to be the centre of the strongest abolitionist sentiment in the country.

Abolitionists like William Lloyd Garrison and Wendell Phillips were New Englanders, and the region was also home to anti-slavery politicians like John Quincy Adams, Charles Sumner, and John P. Hale. When the anti-slavery Republican Party was formed in the 1850s, all of New England, including areas that had previously been strongholds for both the Whig and the Democratic Parties, became strongly Republican, as it would remain until the early 20th century, when immigration would begin to turn the formerly solidly Republican states of Lower New England towards the Democrats.

Aside from the Canadian province of Nova Scotia, or "New Scotland," New England is the only North American region to inherit the name of a kingdom in the British Isles. New England has largely preserved its regional character, especially in its historic places.

Its name is a reminder of the past, as many of the original English-Americans have migrated further west. Today, the region is more ethnically diverse, having seen waves of immigration from Ireland, Quebec, Italy, Portugal, Asia, Latin America, Africa, other parts of the United States, and elsewhere. The enduring European influence can be seen in the region, from use of traffic rotaries to the bilingual French and English towns of northern Vermont, Maine, and New Hampshire, as innocuous as the sprinkled use of British spelling, and as obvious as the region's heavy prevalence of English town and county names, and its unique, often non-erotic coastal dialect reminiscent of southeastern England.

New England is the traditional centre of ethnic English ancestry and culture in the United States. The only place in the U.S. outside New England with a significant majority English ethnicity is Utah-Eastern Idaho, the traditional core of the Jello Belt region, whose proportion of English Americans is actually higher today than that of New England, with Utah being the most English of U.S. States with 29.0% English ancestry, followed by New England states Maine with 21.5% and Vermont with 18.4%.

This population is, in contrast, far more conservative than modern New England and is mainly LDS in religion, but its substratal cultural character is largely reminiscent of both early 19th-century New England and Victorian England (due to later direct handcart immigration).

Geography

New England's long rolling hills, mountains, and jagged coastline are glacial land forms resulting from the retreat of ice sheets approximately 18,000 years ago, during the last glacial period. The coast of the region, extending from southwestern Connecticut to northeastern Maine, is dotted with lakes, hills, swamps, and sandy beaches. Further inland are the Appalachian Mountains, extending through Connecticut, Massachusetts, Vermont, New Hampshire, and Maine. Among them, in the White Mountains of New Hampshire is Mount Washington, which at 1,917 m (6,289 ft), is the highest peak in the northeast United States. It is also the site of the highest recorded wind speed on Earth. Vermont's Green Mountains, which become the Berkshire Hills in western Massachusetts and Connecticut, are smaller than the White Mountains. Valleys in the region include the Connecticut River Valley and the Merrimack Valley.

The longest river is the Connecticut River, which flows from northeastern New Hampshire for 655 km (407 mi), emptying into the Long Island Sound, roughly biscecting the region. Lake Champlain, wedged between Vermont and New York, is the largest lake in the region, followed by Moosehead Lake in Maine and Lake Winnipesaukee in New Hampshire.

Climate

Weather patterns vary throughout the region. Maine, New Hampshire, and Vermont have a humid continental short summer climate, with mild summers and cold winters. Connecticut, Massachusetts, and Rhode Island, have a humid continental long summer climate, with warm summers and cold winters.

Owing to thick deciduous forests, fall in New England brings bright and colourful foliage, which comes earlier than in other regions, attracting tourism by 'leaf peepers'. Springs are generally wet and cloudy. Average rainfall generally ranges from 1,000 to 1,500 mm (40 to 60 in) a year, although the northern parts of Vermont and Maine see slightly less, from 500 to 1,000 mm (20 to 40 in). Snowfall can often exceed 2,500 mm (98 in) annually. As a result, the mountains and ski resorts of Vermont and New Hampshire are popular destinations in the winter.

The lowest recorded temperature in New England was "50 °F ("45.6 °C) at Bloomfield, Vermont, on December 30, 1933. This was tied by Big Black River, Maine in 2009.

Geology

The area is geologically part of the New England province.

Population

In 2005, the total population of New England was 14,239,724 people, roughly a 50% increase from its 1929 population of 9,813,000. If New England were one state, its population would rank 5th in the nation, behind Florida. Its land area, at 62,808.96 sq mi (162,674.5 km), would rank 21st, behind Washington and ahead of Georgia. The region's average population density is 221.66 inhabitants/sq mi (85.59/km²), although a great disparity exists between its northern and southern portions, as noted below. It is much greater than that of the United States as a whole (79.56/sq mi) or even just the contiguous 48 states (94.48/sq mi).

In 2009, two states were among the five highest in the country in divorce rates. Maine was second highest with 13.6% of people over 15 divorced; Vermont was fifth with 12.6% divorced. Connecticut, Massachusetts, and Rhode Island, on the other hand, have below-average divorce rates. Massachusetts is tied with Georgia with the lowest divorce rate in the U.S., at 2.4%.

Southern New England

Three-quarters of the population of New England and most of the major cities are in the states of Connecticut, Massachusetts, and Rhode Island. Their combined population density is 786.83/sq mi, compared to northern New England's 63.56/sq mi (2000 census). The most populous state is Massachusetts, and the most populous city is Massachusetts' political and cultural capital, Boston.

Coastal New England

The coastline is more urban than western New England, which is typically rural, even in urban states like Massachusetts. This characteristic of the region's population is due mainly to historical factors; the original colonists settled mostly on the coastline of Massachusetts Bay.

The only New England state without access to the Atlantic Ocean, Vermont, is also the least urbanized. After nearly 400 years, the region still maintains, for the most part, its historical population layout.

New England's coast is dotted with urban centres, such as Portland, Portsmouth, Boston, New Bedford, Fall River, Providence, New Haven, Bridgeport, and Stamford as well as smaller cities, like Newburyport, Gloucester, Biddeford, Bath, Rockland, Newport, Westerly, Rhode Island, and the small twin cities of Groton, Connecticut and New London.

Urban New England

Southern New England forms an integral part of the Bos Wash megalopolis, a conglomeration of urban centres that spans from Boston to Washington, D.C..

The region includes three of the four most densely populated states in the United States; only New Jersey has a higher population density than the states of Rhode Island, Massachusetts, and Connecticut.

Greater Boston, which includes parts of southern New Hampshire, has a total population of approximately 4.4 million,, while over half the population of New England falls inside Boston's Combined Statistical Area of over 7.4 million. The most populous cities are as of 2000 Census (2008 estimates in parenthesis):

1. Boston, Massachusetts: 589,141 (620,535)
2. Providence, Rhode Island: 173,618 (171,557)
3. Worcester, Massachusetts: 172,648 (182,596)
4. Springfield, Massachusetts: 152,082 (150,640)
5. Bridgeport, Connecticut: 139,529 (136,405)
6. Hartford, Connecticut: 124,558 (124,062)
7. New Haven, Connecticut: 123,626 (123,669)
8. Stamford, Connecticut: 117,083 (119,303)
9. Waterbury, Connecticut: 107,271 (107,037)
10. Manchester, New Hampshire: 107,006 (108,586)
11. Lowell, Massachusetts: 105,167 (103,615)
12. Cambridge, Massachusetts: 101,355 (105,596)

During the 20th century, urban expansion in regions surrounding New York City has become an important economic influence on neighbouring Connecticut, parts of which belong to the New York Metropolitan Area.

The US Census Bureau groups Fairfield, New Haven and Litchfield counties in western Connecticut together with New York City, and other parts of New York and New Jersey as a combined statistical area.

Economy

Several factors contribute to the uniqueness of the New England economy. The region is geographically isolated from the rest of the United States, and is relatively small. It has a climate and a supply of natural resources (such as granite, lobster, and codfish) that are different from many other parts of the country.

Its population is concentrated on the coast and in its southern states, and its residents have a strong regional identity. America's textile industry began along the Blackstone River with the Slater Mill at Pawtucket, Rhode Island, and was duplicated at similar sources of water power such as Woonsocket, Rhode Island, Uxbridge, Massachusetts, and Lawrence, Massachusetts.

Exports consist mostly of industrial products, including specialized machines and weaponry (aircraft and missiles especially), built by the region's educated workforce. About half of the region's exports consist of industrial and commercial machinery, such as computers and electronic and electrical equipment. This, when combined with instruments, chemicals, and transportation equipment, makes up about three-quarters of the region's exports.

Granite is quarried at Barre, Vermont, guns made at Springfield, Massachusetts and Saco, Maine, boats at Groton, Connecticut and Bath, Maine, and hand tools at Turners Falls, Massachusetts. Insurance is a driving force in and around Hartford, Connecticut. New England also exports food products, ranging from fish to lobster, cranberries, Maine potatoes, and maple syrup. The service industry is also highly important, including tourism, education, financial and insurance services, plus architectural, building, and construction services. The U.S. Department of Commerce has called the New England economy a microcosm for the entire United States economy.

As of December 2008, the unemployment rate in New England was 6.9%, below the national average. New Hampshire, with the lowest of the six states, had a rate of 4.6%. The highest was Rhode Island, with 10.0%. The metropolitan statistical area (MSA) with the lowest rate, 2.5%, was Manchester, New Hampshire; the MSA with the highest rate, 10.8%, was Lawrence-Methuen-Salem, in Massachusetts and southern New Hampshire.

New England has two of the ten poorest cities (by percentage living below the poverty line) in the United States: the state capital cities of Providence, Rhode Island and Hartford, Connecticut. These cities have struggled as manufacturing, their traditional economic mainstay, has declined.

On the other hand, New Hampshire and Connecticut had some of the lowest poverty rates in the country in 2006.

With its rocky soil and climate, New England is not a strong agricultural region. Some New England states, however, are ranked highly among U.S. States for particular areas of production. Maine is ranked ninth for aquaculture, and has abundant potato fields in its northeast part. Vermont fifteenth for dairy products, and Connecticut and Massachusetts seventh and eleventh for tobacco, respectively. Cranberries are grown in Massachusetts' Cape Cod-Plymouth-South Shore area, and blueberries in Maine. As of 2007, the inflation-adjusted combined GSPs of the six states of New England was $744.6 billion, with Massachusetts contributing the most, and Vermont the least.

Energy

The region is mostly very energy efficient compared to the country at large. Rhode Island has the lowest per capita energy consumption of any state in the country and five of the New England states placed in the lowest eleven. Maine, by contrast, had the 17th-highest per capita consumption.

Maine is leading New England in wind power production, and the state's agenda states that by 2030, the state will produce twice the amount of energy in wind power than it consumes.

The six New England states collectively have the highest electricity costs in the nation. The best rates are in Vermont which stands 41st in the country; the worst, Rhode Island, is 50th (out of 51).

Three of the six New England states are among the country's highest consumers of nuclear power: Vermont (first, 73.7%), Connecticut (fourth, 48.9%), and New Hampshire (sixth, 46%).

Agriculture

In the US, milk prices collapsed in 2009. Vermont Senator Bernie Sanders accused Dean Foods of controlling 70% of New England's milk market. He has requested the United States Department of Justice to pursue an anti-trust investigation.

Infrastructure

Six mainline Interstate highways cross New England, with at least one serving each state and its respective capital city:

Interstate 84 enters New England at Danbury, Connecticut, and crosses that state to the northeast; connecting the city of Waterbury and the state capital of Hartford before terminating at a junction with Interstate 90 in Massachusetts.

Interstate 90, also signed east-west, carries the Massachusetts Turnpike designation as it crosses the state. I-90 enters Massachusetts at West Stockbridge and travels eastward to its terminus in Boston; connecting the cities of Springfield and Worcester and intersecting many of New England's major north-south routes.

Interstate 89, signed north-south, begins at a junction with Interstate 93 just south of Concord, New Hampshire. I-89 travels to the northwest towards its terminus at the Canadian border, connecting Lebanon, the state capital of Montpelier, and Burlington (Vermont's largest city) along the way.

Interstate 91 begins in New Haven, Connecticut at a junction with Interstate 95, running north from there throughout Connecticut, Massachusetts, and Vermont until it reaches the Canadian border. I-91 parallels U.S. Route 5 for its entire length, and much of the route also follows the Connecticut River, linking many of the major cities and towns along the river including Hartford, Springfield, and Brattleboro. I-91 is the only Interstate route within New England that intersects all five of the others.

Interstate 93 begins in Canton, Massachusetts at a junction with Interstate 95, running northeast ward from there through the city of Boston. I-93 travels north from Boston and into New Hampshire, where it serves as the main Interstate highway through that state and links many of the larger cities and towns (including the capital, Concord, and Manchester). I-93 eventually enters Vermont and reaches its northern terminus at a junction with Interstate 91.

Interstate 95, which runs along the East Coast, enters New England at Greenwich, Connecticut, and runs in a general northeastern direction along the Atlantic Ocean, eventually heading through Maine's sparsely-populated north country to its northern terminus at the Canadian border. I-95 serves many of the coastline's cities, including the state capitals of Providence and Augusta, while serving as a partial beltway around Boston. I-95 travels through every New England state except Vermont, and is the only two-digit Interstate highway to enter the states of Rhode Island and Maine.

The Massachusetts Bay Transportation Authority (MBTA) provides rail and subway service within the Boston metropolitan area, bus service in Greater Boston, and commuter rail service throughout Eastern Massachusetts and parts of Rhode Island.

Amtrak provides rail service throughout New England. Boston is the northern terminus of the Northeast Corridor line.

Bibliography

Belden C.: *Landscapes of the Sacred: Geography and Narrative in American Spirituality*, New York: Paulist Press, 1988.

Berry, J.K. : *Beyond Mapping, Concepts, Algorithms and Issues in GIS*, Fort Collins, CO: GIS World Books, 1993.

Bolstad, P. : *GIS Fundamentals, A first text on Geographic Information Systems, Second Edition*, White Bear Lake, MN, Eider Press, 2005.

Cronon, William: *Changes in the Land: Indians, Colonists, and the Ecology of New England*. New York: Hill and Wang, 1983.

Davis, Mike: *Ecology of Fear: Los Angeles and the Imagination of Disaster*. New York: Vintage, 1999.

Elangovan K. : *GIS, Fundamentals, Applications and Population Geography*, New India Publishing Agency, New Delhi 2006.

Harvey, David: *Explanation in Geography*. London: Edward Arnold, 1969.

Harvey, Francis : *A Primer of GIS, Fundamental Geographic and Cartographic Concepts,* The Guilford Press, 2008.

Jackson, Peter: *Maps of Meaning: An Introduction to Cultural Geography*. London and Boston: Unwin Hyman, 1989.

James S. Duncan: *Writing Worlds: Discourse, Text, and Metaphor in the Representation of Landscape*. London and New York: Routledge, 1992.

James, T.: *The Best Poor Man's Country: A Geographical Study of Early Southeastern Pennsylvania*. Baltimore: Johns Hopkins Press, 1972.

King, Anthony D.: *Buildings and Society: Essays on the Social Development of the Built Enviroment*. London: Routledge & Kegan Paul, 1980.

Lewis, Peirce. *New Orleans: The Making of an Urban Landscape.* Cambridge, MA: Ballinger Publishing Co., 1976. Annotation

Liam Kennedy: *Urban Space and Representation.* Sterling, VA: Pluto Press, 2000.

Lung, Y., and van Tulder, R. : *Cars, Carriers of Regionalism*, Palgrave Macmillan, New York, 2004.

McDonnell, R.A. : *Principles of Geographical Information Systems*, Oxford University Press, Oxford, 1998.

Michael P.: *The Making of the American Landscape.* Boston: Unwin Hyman, 1990.

Michael, H.: *City Form and Natural Process: Towards a New Urban Vernacular.* London: Croom Helm, 1984.

Peter, G.: *The Slow Plague: A Geography of the AIDS Pandemic.* Oxford and Cambridge, MA: Blackwell Publishers, 1993.

Ray Allen: *The Far Western Frontier, 1830-1860.* The New American Nation Series. New York: Harper, 1956.

Richard SCASE : *Managing Creativity,* Open University Press, 2000.

Stephen, D.: *Fields of Vision: Landscape Imagery and National Identity in England and the United States.* Cambridge: Polity, 1993.

Tulder, R. : *Cars, Carriers of Regionalism*, Palgrave Macmillan, New York, 2004.

Victor, A. : *Geography of Art and Culture*, Contributions to Economic Analysis, Elsevier, 2004.

Watkins, Eric: *The Middle Eastern Environment*, London, The British Society for Middle Eastern Studie, 1995.

William G. BOWEN : *Performing Arts, The Economic Dilemma,* Thousand Oaks, SAGE, London, 2002.

Worboys, Michael, and Matt Duckham. : *GIS, a Computing Perspective*, Boca Raton, CRC Press, 2004.

Index

E

F

G

H

I

J

L

□□□